BIM in Principle and in Practice

Institution of Civil Engineers

publishing

BIM in Principle and in Practice

Peter Barnes and Nigel Davies

Published by ICE Publishing, One Great George Street, Westminster, London SW1P 3AA.

Full details of ICE Publishing sales representatives and distributors can be found at:
www.icevirtuallibrary.com/info/printbooksales

Other titles by ICE Publishing:

Intelligent Buildings: Design, Management and Operation (2nd ed).
Derek Clements-Croome. ISBN 978-0-7277-5734-0
Asset Management: Whole-Life Management of Physical Assets.
Chris Lloyd. ISBN 978-0-7277-3653-6
Sustainable Infrastructure: Principles into Practice.
Charles Ainger and Richard Fenner. ISBN 978-0-7277-5754-8

www.icevirtuallibrary.com

A catalogue record for this book is available from the British Library

ISBN 978-0-7277-5863-7

© Thomas Telford Limited 2014

ICE Publishing is a division of Thomas Telford Ltd, a wholly-owned subsidiary of the Institution of Civil Engineers (ICE).

Commissioning Editor: Jennifer Saines
Production Editor: Vikarn Chowdhary
Market Specialist: Catherine de Gatacre
Cover image: Courtesy of Shutterstock.com ©Sergey150770

Typeset by Academic + Technical, Bristol
Index created by Pauline Davies
Printed and bound by CPI Group (UK) Ltd, Croydon CR0 4YY

FSC
www.fsc.org
MIX
Paper from
responsible sources
FSC® C013604

Contents

Preface

Building information modelling (BIM) is the topic of the moment. Already many companies have experienced the benefits (and some of the pitfalls) of BIM while implementing BIM technology and the processes that it enables.

However, there is still a large portion of the construction and civil engineering industries that has yet to take the first steps into the world of BIM. Furthermore, and to very many people, BIM is a mystery that is expressed in a foreign language. It is perceived as being highly technical and difficult to grasp as a concept.

Despite that, those same people know that BIM cannot be ignored, and it is apparent that the consultants, contractors and subcontractors that get involved with BIM now, or at the least inform themselves about what BIM can do for their practice or company and how a BIM-enabled company might better serve the industry, will soon be in high demand.

As BIM becomes more commonly used, and as it is specified more often in contracts, more benefits will surface. BIM will reshape the construction industry. It is not a question of if; it is only a question of when.

Those who feel that the boat is doing just fine and should not be rocked may find themselves scrambling for BIM tools and rushing into perhaps ill-advised choices once BIM becomes a general requirement, be it for economic, environmental or other reasons.

The purpose of this book, therefore, is to explain in relatively simple terms what BIM is and how it can be applied in practice. With this aim in mind, the early chapters of this book deal with the theory of BIM, while the later chapters provide examples of BIM used in practice.

BIM is a fascinating subject and offers exciting opportunities. Apart from the obvious benefits of design efficiencies and the contribution towards sustainable and green building design, the areas in which BIM may have even greater long-term effect areas are those of collaboration, partnering, contract relationships, the resurgence of terotechnology in the modern world, and even dispute resolution.

The authors of this book have a particular interest in BIM, and realise that, because of the importance of BIM to the construction and civil engineering industries, the benefits and potential gains of using BIM need to be widely broadcast, and this is the purpose of this book. It is hoped that the readers of this book will find it to informative and interesting, and that it will develop an interest in them to investigate BIM further and apply BIM techniques in the coming years.

Finally, we are both most grateful to all those who have contributed to our knowledge about the subject matter of this book over the years, and we give special thanks to our families who have been a great support to us during the writing of this book.

About the authors

Peter Barnes

Peter Barnes is a Director of Blue Sky ADR Ltd. He has been actively involved in the construction industry for almost 40 years, and over that time has been actively involved in the design process, contracting and consultancy works in the construction and civil engineering industries.

Peter is a Chartered Conservationist, and through his involvement with the Joint Contracts Tribunal (where he is a member of the Council) he has developed a great interest in sustainable building processes and collaborative contract relationships. This interest led to his decision to co-author this book on BIM in principle and in practice.

Nigel Davies

Nigel Davies is a Director of Blue Sky ADR Ltd. He has a wide and extensive range of practical and commercial experience. Among other things, Nigel specialises in providing commercial and contractual management, construction claims and alternative dispute-resolution services to the construction and civil engineering industries. Nigel has worked and acted for employers, architects, main contractors and specialist subcontractors in providing quantity surveying and construction contract services.

Nigel sits on the Council of the Joint Contracts Tribunal and is a member of the Sustainability Sub-Committee of that council. It is because of these roles that the use and integration of the BIM process in this and other regards is of particular interest to Nigel.

BIM in Principle and in Practice
ISBN 978-0-7277-5863-7

Chapter 1
Introduction

There can be little doubt that BIM is here to stay, but there is certainly a good deal of confusion about what exactly BIM is. BIM stands for 'building information modelling' or 'building information management'; and, because of this, sometimes the acronym BIM(M) is used, which stands for 'building information modelling and management'.

At its most basic level, BIM is a computerised process that is used to design, understand and demonstrate the key physical and functional characteristics of a building (or a construction or civil engineering project) on a 'virtual' computerised model basis. BIM therefore provides the opportunity to concurrently design and visualise the building in 3D, although it needs to be understood that the opportunity to visualise in 3D is a product of BIM, and not the process of BIM itself.

At its more advanced levels, BIM is the use of a computer software model to simulate the construction and operation of a building, and use of BIM at these levels is something that is bound to happen more and more in the future. The resulting model, a 'building information model', is a digital representation of the building, from which views and data appropriate to various users' needs can be extracted and analysed to generate information that can be used to make decisions and improve both the process of delivering the building and the entire life-cycle use of the building.

While the use of BIM appears to be most appropriate for new buildings, it is not only for new buildings, but rather it is for all buildings, both old and new. Therefore, in the future, the retrospective modelling of existing building stock represents a significant opportunity for the effective utilisation of BIM and for, among other things, the attainment of significant efficiency gains.

The UK Government Construction Strategy published in 2011 required that fully collaborative 3D BIM should be used on all centrally procured UK Government construction projects by 2016. Therefore, while BIM is, in any event, organically growing in use, the UK Government's Construction Strategy will no doubt advance greatly the use of BIM in the UK over the next few years, as will the need to improve efficiency and reduce costs

1

in both the public and the private sector of the construction and civil engineering industries.

When applied correctly, BIM is intended to make substantial cost savings throughout the whole life cycle of a building, from design, through construction and maintenance, to regeneration and eventual disposal or recycling.

With respect to the design and construction phase, the BIM process, when fully applied, possesses the potential to save valuable resources, including time, money and natural materials. BIM will do this by reducing the amount of inaccurate and conflicting information, and this, in consequence, will reduce variations, alterations and delays. These benefits will be achieved because the building is constructed in the 'virtual' world before it is built physically.

The adoption of the BIM process during the design and construction phase of a building by those parties interested in the successful completion and outcome of the building, from initial design to practical completion of the property, offers the opportunity to achieve the systematic coordination and use of all data available.

Over recent years, a digital revolution has occurred across many industries, changing the way we live. No more than 20 years ago or so people could generally only buy music and films from a shop and they could only buy books from a book shop. Now, of course, this is no longer the case. iPods, MP3 players, Kindles, iPads, personal computers and the internet have changed the way we live our lives.

Information technology (IT) has long since become part of our everyday lives, both at home and in the workplace, and the BIM process is another, construction-specific chapter in the tailoring of IT to the specific needs of the construction and civil engineering industries.

Whereas, previously, documents such as drawings, schedules and specifications were written and stored in hard copy form, this information is now digitally generated and stored from and within BIM models. Within the construction and civil engineering industries we are now on the threshold of seeing digital information flows from inception through to demolition of projects, which is bound to result in large efficiencies being achieved in many areas.

Another major aspect of BIM is the potential full collaboration of the entire project team – the employer, the architect, the engineers, the consultants, the contractor and the specialist contractors – in developing the project design. This full collaboration not only allows for increased speed of project delivery, enhanced economics for the project and true lean construction all at levels but also has the potential to change the relationships between the participants in the construction industry, from the more traditional contracts based on obligations and rights to the more modern partnering associations based on a fair allocation and sharing of risks and liabilities.

This book is intended to assist in the understanding of BIM – how BIM can be used now, and how it may be developed further in the future – and also provides answers to many of the common questions asked about BIM. While it is the case that BIM is still very much in its infancy in many respects, and there are very few fully BIM-compliant projects at the moment, case studies and templates are provided in this book to show how the BIM process (or elements of the BIM process), as described in principle, may actually be applied in practice.

BIM in Principle and in Practice
ISBN 978-0-7277-5863-7

ICE Publishing: All rights reserved
http://dx.doi.org/10.1680/bimpp.58637.005

Chapter 2
An overview of BIM

2.1   What is BIM?

Most people consider that BIM stands for 'building information modelling' (although some consider that it stands for 'building information management'). Because of this, sometimes the acronym BIM(M) is used, which stands for 'building information modelling and management'. However, it does not really matter that much what the acronym BIM actually stands for in terms of the words used, the important thing to understand is 'What is BIM'?

With that in mind, and particularly in the context of the construction and civil engineering industries, BIM is a process that relies on a computerised virtual 3D model of a building (or a construction or civil engineering project or some other facility, but which, for the ease of future reference, will simply be referred to as a 'building') which reacts to changes in the same way that the actually constructed building would.

This reaction is achieved by applying to a computer model software applications that are used to design, understand and demonstrate the key physical and functional characteristics of the building throughout its entire life cycle (i.e. from its conception to its final demolition and/or disposal). The BIM process is underpinned by the creation, collation and exchange of shared 3D submodels (i.e. models of a part of the building or of part of the works) with reactive and structured data attached to them. Because this process is one of design and construction being carried out in the virtual world, BIM is sometimes also referred to as 'virtual design and construction' (VDC).

2.2   The background and history of BIM

To consider the background and history of BIM we need to start with computer aided design (CAD), which emerged in the 1950s and 1960s within academia and some very large manufacturing industries that were able to afford mainframe computing, something that was a relatively rare occurrence at that time. Each of those organisations normally produced their own software to produce 2D drawings by computer, and the commercial return of those CAD systems for those organisations lay in the time saved

5

as compared with the time that would normally be spent preparing 2D drawings in the traditional way of drafting by hand.

As time went on, CAD became more and more widely used, and because of that increased commercial use there was obviously an concomitant development of the software that CAD was based on. This software evolved such that CAD eventually had the ability to communicate product design information directly from the drawing board to computer-driven manufacturing tools. In other words, products could be manufactured straight from the computer without going through any further pre-production stages. However, CAD was principally still dealing with 2D drawings.

The next step was to move from 2D drawings to 3D models, and the principal difficulty that was faced in this transition was how to mathematically represent intricate objects and surfaces within a computer model such that they would be recognised by the software produced for this purpose.

Surface modelling is complex, and before the advent of computers and advanced software there was no way to undertake numerical computations other than by hand. In the 1940s, aircraft designers, for example, would use a technique that had been used for many years in ship building for hull production, by which they would model aircraft design using thin wooden strips called 'splines' to pass through fixed points. The splines were held in place using lead weights. The aircraft model design together with the fixed points would then be plotted on paper to full size, a process which was referred to as 'lofting'.

It was the development of an electronic version of splines and lofting that was used in CAD systems in the transition from 2D plans to 3D modelling. Therefore, to plot a curve, the mathematical equation of the curve needed to be established by the software along the entire length of the curve. Other than for the most mundane straight lines and conic shapes, the surfaces of most manufactured products are actually the result of an enormously complex mathematical formula, and it was this issue that the software needed to deal with.

The early CAD was developed further during the 1960s and early 1970s, and although the use of CAD was made more accessible through the use of smaller mini-computers, which made the process more economical, the cost of using CAD was still relatively high and in many cases prohibitive in relation to the overall cost of the project.

However, during the late 1970s there was rapid commercial growth in what was becoming the CAD industry, and this was stimulated by the increased availability of improving, and less expensive, hardware technology combined with major advances in software development. These advances, particularly in respect of surface and solid modelling, offered manufacturing industries the possibility of enhancing their computer-aided design and manufacturing systems; and because of this, two main methods of displaying and recording shape information began to appear in the 1970s and 1980s – constructive solid geometry and boundary representation.

The rate of progression of CAD continued to increase during the 1980s, first with the arrival of much less expensive forms of mini-workstation computers, and then in the following years through the increased use of personal computers (PCs). It was largely because of the arrival and the proliferation of the use of PCs that CAD was given a major push forward.

The increased use of CAD contributed to the breakdown of buildings into their constituent parts, which then developed into library databases being produced for individual products and components. *Uniclass: The Unified Classification for the Construction Industry* (Construction Project Information Committee, 1997) was first published in 1997 and was based on ISO/TR 14177:1994, Classification of information in the construction industry (International Organisation for Standardisation, 1994). Uniclass was a new classification scheme for the construction industry in the UK that organised and structured a library of materials and products literature together with other project information.

Around about the same time, in the early 1990s software was developed that integrated graphical analysis and simulations to provide information about how a building might perform under different conditions, including the building's orientation, geometry, material properties and building systems. Soon after this, software that utilised a constraint-based interactive (i.e. parametric) modelling engine was being developed. This software meant that one part of a model would change if another related element of the model was changed. In another related area in the late 1980s, software was being developed that allowed for time-based phasing of the construction process to be represented in a 3D virtual model.

By the beginning of this century, all these software developments were being brought together, and this represented a major advance in the world of software modelling. This more advanced software created a virtual platform that utilised a 3D model together with interactive/parametric 'family' submodels, which allowed for a time attribute to be added to a component to allow a fourth dimension (4D) of time to be linked into the base model.

Since then the software has been developed further, and this has contributed to the growth of BIM and, as a result, models with several other dimensions (e.g. cost, energy) have been developed ever more rapidly over the past few years.

When considering the future of BIM, it must be remembered that, when CAD was first introduced, it was viewed as a tool for very large companies and for very large projects. However, now it is used as a standard tool across the construction and civil engineering industries, and very rarely is the manual drafting of drawings undertaken. Therefore, based on this relatively recent history of a similar technology, it should be apparent that history will almost certainly repeat itself, and BIM will also similarly become a standard platform and tool within a very few years.

2.3   Why does BIM matter?

We have been producing buildings for hundreds of years without the use of BIM, so why does the use of BIM matter at all? At the very basic level, BIM represents technological progress, and in 'philosophical' terms humankind (and, at a less esoteric level, the construction and civil engineering industries) thrives and develops on progress.

However, at a much more mundane and less philosophical level, BIM brings with it many benefits, and as it develops further it will bring many more benefits that can only help to make the construction and civil engineering industries become more efficient and effective. Some of the more important benefits that the BIM process brings, or will potentially bring in the future, are listed below. BIM:

- permits everyone involved in the process (including the client) to have a clear understanding of the concept design by providing the ability to visualise (by way of a virtual 3D model walkthrough) what is actually to be built
- helps to provide a very useful and successful marketing tool for all those involved in a project
- assists with better design and space planning to optimise the use of the building
- allows for the remodelling of alternative layouts and options, and for the integration of a selection of different construction techniques
- permits value engineering exercises to be carried out and projected on a virtual basis
- allows for the full coordination of the various consultants' designs, and helps prevent conflict between those designs
- provides the ability to identify collisions and clashes (e.g., identifying ductwork running into structural members) before construction starts on site
- encourages collaboration of working and design between the various consultants
- provides the ability to use information in the 3D model to directly create fabrication drawings, thus avoiding what can often be a problematic and difficult step in the construction process
- facilitates fewer errors and corrections on site, and therefore leads to less rectification and remedial works, which results in lower overall costs

- permits a higher reliability of expected site conditions, which allows the opportunity for more prefabrication of components off-site, which will usually be at a higher quality and/or at a lower cost and which will have the effect of reducing waste
- allows for evaluation during the early design phases of the energy expected to be used over the life cycle of the project, and this can improve the carbon footprint of the facility
- permits contractors to carry out a dry run of the proposed construction
- assists with the ability to do more 'what if' scenarios, such as looking at various sequencing options, site logistics, hoisting alternatives, etc., and this can result in safer working conditions
- gives the opportunity for programme visualisation and makes it easier to understand the dependency that various activities have upon each other
- allows for material 'take-offs' to be produced automatically
- can accurately and rapidly generate an array of essential estimating information, such as materials quantities and costs, size and area estimates, costs and productivity projections
- permits materials quantities and costs, size and area estimates, costs and productivity projections to be carried out as changes are made and/or variations are introduced
- makes it simpler to produce 'as built' drawings upon completion
- assists with maintenance, facilities management and life-cycle replacement issues
- can be used to test the optimisation of building maintenance, renovations and energy efficiency, as well as to monitor life-cycle costs.

Of course, most of the above tasks have previously been undertaken without BIM being used. However, one of the great advantages of BIM is the ease and speed with which the functions may be tested and applied on a virtual basis before any actual construction work commences at all.

In reality, it is only the relatively recent advances in technology that have allowed such virtual design to be carried out, and the construction and civil engineering industries clearly need to make use of such modern technology to be effective and efficient in the modern world.

In addition to the more obvious benefits with respect to the construction process, as outlined above, BIM will bring additional opportunities in the form of exploiting existing and new export markets for BIM 'know how' and BIM-ready products; and the additional and collateral effect of the widespread adoption of BIM both domestically and internationally should bring with it a great opportunity for niche software developers and consultants. Therefore, in overview terms, BIM potentially enables collaboration between clients, designers and contractors in technological ways (and possibly through contracts etc.) that have not been available to the construction and civil engineering industries in the past.

Furthermore, and irrespective of anything else, and in the most basic of terms, you simply cannot stand in the way of progress. By way of an analogy, we could still carry out calculations using a pen and paper approach (or even an abacus!), but the technology available (e.g. calculators, computers) has made that approach essentially redundant; and it is inevitable that before too much longer the use of design without BIM on anything other than the most basic of projects will be considered to be equally inefficient and outdated.

One major point that many people are perhaps overly concerned about is 'How does BIM actually work?', whereas what they should really be interested in is 'What can BIM do for the construction and civil engineering industries?'. In other words, we really should be more interested with the outcome of the process rather than the process itself.

For example, relatively few of us understand how modern technology really works – most of us do not know how the internet or the world wide web works, but that does not stop us from using those facilities virtually every day of our lives as a platform or as a tool. In fact, for many of us, we would not be able to function properly in our business lives without access to the internet, email etc.

So, the most important thing about BIM is to understand that it has the capacity to make the construction and civil engineering industries more efficient and effective. We do not really need to understand the advanced software and the complexities of BIM which will, in effect, and using an analogy of a high performance car, be 'hidden under the bonnet'.

In this way, if you consider how user friendly a tool such as an internet map search engine is, where, with a few spins of the wheel and a click of the computer mouse it is possible to zoom in from a small-scale map of Europe to find the telephone number of a brick supplier in Nottinghamshire, the software technology and the vast amount of 'big data' behind the user interface is simply mind boggling. However, we do not need to know about those technological details, we just need to know how to use the tool.

This is now something that is paralleled in BIM, and will be even more so through the increased usage of the BIM technology. Therefore, when operating the BIM process, a user may, for example, view the building, then navigate through the floors to the rooms, find the system or elements of the work that they are interested in and then review the embedded specification.

Therefore, in this regard, BIM should be viewed more as a process or a tool for carrying out information modelling and information management in a team environment, rather than as being a technical skill that needs to be acquired. There will be plenty of software and hardware providers that will deal with the technical nuts and bolts, which will leave those practitioners in the construction and civil engineering industries who will use BIM in their everyday life to concentrate on the operation of and the benefits to be gained from using BIM.

While BIM is certainly not the solution to all problems, the rich 3D experience, digital simulations, rehearsals of all stages of the design, build and operate processes, and the information embedded within the BIM models will facilitate well-informed decision-making, which should result in better business outcomes, greater clarity, improved communication, less risk and, ultimately, better efficiency. Clearly BIM is an evolving concept, and it is anticipated that the list of benefits and potential benefits noted above will become more apparent through increased usage.

However, the ultimate aims of BIM are to improve efficiencies, allow for leaner and 'greener' buildings, and achieve cost savings in the construction process and the life-cycle costing of a building project, and it is intended to make an integrated and collaborative approach to construction possible. It is for all these reasons that BIM matters.

2.4 What is FIM?

As more and more buildings are created with the use of BIM, this will encourage the integration of BIM into facilities management. This integration is known as facilities information management or facilities information modelling (FIM).

If the BIM model is properly maintained during construction, it will become a tool that can be used by the building owner or user to manage and operate the facility. Therefore, proposed modifications and upgrades will be able to be evaluated for cost effectiveness, and data contained in the BIM model can be used to manage maintenance.

Through BIM, data can be extracted on items such as room sizes, equipment data and room finishes, from conception and throughout the entire life cycle of the building. Also, upon completion of the construction of a project, information that previously and typically would have been gathered by people being sent to site will be gathered from the BIM model instead, and this will naturally offer significant time and cost savings.

By integrating BIM into facilities management tools (such as computerised maintenance management systems (CMMS) and computer aided facilities management (CAFM)), a visual and interactive environment is added to the workflow. For example, an air handling unit could have warranty information embedded within the BIM model, and this could be linked to operation and maintenance manuals, to real-time data such as air flow and temperature, and also to wiring diagrams and spare parts lists. All this information and data being available in one location would make the maintenance and/or repair of the air handling unit much easier to control and manage.

In addition, model element location information could be linked with CMMS and CAFM systems, and this would allow facilities management teams to have the necessary information about maintenance and other issues in a more tangible form and at their fingertips, through the model, which would create efficiencies in their day-to-day tasks.

When the FIM system is operating fully, it is anticipated that a facilities manager will not even need to know how to navigate a model. Through the use of bar codes, real-time location systems and portable tablet devices, the facilities manager would simply need to scan codes located in the rooms of the building and the model view would automatically orientate to match the location to which the tablet device was being pointed. This could be taken a step further by accessing different layers of the model, essentially creating an interactive 'x-ray' portal into walls.

Where a model is created by the designer and then updated throughout the construction phase, it would have the capacity to become an 'as-built' model, which also could be turned over to the owner and/or the end user. This model would be able to contain all the specifications, operation and maintenance (O&M) manuals and warranty information that would be required for future maintenance and/or repair. This approach would clearly eliminate the problems that can currently be experienced if the O&M manuals have been misplaced or are kept at a remote or different location from where they need to be.

Furthermore, sensors in the building could feed back and record data relevant to the operation phase of a building, enabling the BIM model to be used to evaluate energy efficiency, monitor a building's life-cycle costs and optimise its energy and cost-efficiency. It would also enable the owner to evaluate the cost-effectiveness of any proposed upgrades to the systems. There is even research presently being undertaken at Princeton University in the USA to embed ultrathin radios in plastic sheets. These will be applied to structures to develop a flexible structural health monitoring system for bridges, buildings and other structures, and the radios in the sheets will feed back data from a large area of the structure (rather than isolated spots as is presently possible from individual sensors) if unexpected stresses or strains are experienced within the structure. Therefore, one day in the future, an overweight load may be driving across a bridge, and the radio sensors may say 'Get off – you will break my back!' – but that may be a step too far.

Of course, to be truly effective, facilities management professionals will need to engage with the initial design process and also with the developing BIM process if the full potential that BIM holds for delivering value and cost savings over a building's entire life cycle is to be realised. Facilities managers therefore need to be involved at the design and specification stage if their input is to be fully utilised, and they should become involved in the development of standards, systems of classification and data sets, ensuring that the required technology is developed in a way that is useful for them and for the owner/end user.

Despite this, the knowledge and expertise of the facilities management profession should not only be utilised at the design stage but in the development of the technology for facilities management purposes, as it is facility managers who have experience and knowledge of lifetime value and whole-life costings, and who bring the insight and understanding of the end user of the buildings.

Essentially, when operating fully, BIM can help create and maintain buildings that are more efficient, have lower carbon emissions, cost less to run and are better, more effective and safer places to live and work.

2.5   Is the construction industry BIM/FIM ready?

The answer to the question 'Is the construction industry BIM/FIM ready yet?' is probably 'No' or at least 'Not yet'. Clearly, BIM is still, in many ways, very much in the innovation stage, but the construction and civil engineering industries generally recognise the need for BIM and the potential benefits that BIM will bring. Of course, there is a major difference between recognising the need and potential benefits of BIM and actually applying it in practice.

Clearly there are certain barriers (or perceived barriers) that may prevent organisations from fully engaging with BIM. These barriers include the view that:

■ the market is not ready and BIM is still in the innovation stage, and it is therefore too early to get involved with BIM
■ training costs are high, the education requirements are unknown and the learning curve is steep
■ an investment in new technology, hardware and software is needed, and the potential cost of that investment is not justified by the potential savings that may be gained at this stage.

We now look at these issues in turn.

2.5.1   The market is not ready and BIM is still in the innovation stage, and it is therefore too early to get involved with BIM

While it is true that BIM is still being developed, it is being used by more and more design consultants and specialist contractors. Bearing in mind that the UK Government

Construction Strategy of 2011 requires that all central UK Government construction procurement will have to be fully collaborative 3D BIM (with all project and asset information, documentation and data being electronic) as a minimum by 2016, it is certainly not too early to get involved, and the time to get on the BIM train is now, otherwise participants will be running after that train after it has left the station.

2.5.2 Training costs are high, the education requirements are not known and the learning curve is steep

This general statement is true of nearly all new innovations, but is really a very short-term view to hold. If applied and operated as envisaged, BIM will give an ongoing saving over the long term, and any initial training costs should soon be absorbed into the overall cost savings of an organisation. Therefore, likely future participants in the BIM process need to concentrate on raising the awareness of BIM by their staff, and to ensure that the appropriate training opportunities are pursued as they become available. Most organisations will become involved with BIM over the next few years, even if it is only in supplying and managing information. Therefore, at the very least, an understanding of the key reference points, COBie UK 2012 (Construction Operations Building Information Exchange) and PAS 1192-2:2012 (BIM Task Force, 2012a,b) (matters that are dealt with in more detail later within this book) should be acquired.

While the UK Government's BIM Task Group has not recommended the introduction of a standard or accreditation system for BIM training and education, it has produced a description of the learning outcomes that BIM training and education courses should consider. This 'learning outcomes framework' is being tested with commercial training providers, professional institutions and academia, and will be used to inform clients and the supply chain of the group's analysis and selection of the appropriate staff training programmes, and will be used by training providers to introduce a broad coverage of BIM-related courses across strategic, management and technical roles. These courses will be required to increase the level of skill in the construction and civil engineering industries in order to support the UK Government's collaborative 3D BIM ambition for 2016.

Currently, professionals are trained and educated within their own professional boundaries. However, BIM and the collaborative working arising through the use of BIM will require new skill sets which can only really be achieved if there is some cross-institutional cooperation on formulating curricula that cross over the existing boundaries between the professions.

Of course, architects, other designers, contractors and suppliers have deeply embedded working practices and long-standing delineations of professional responsibility and liability. However, if BIM is to be successful in fulfilling the UK Government's aspirations, project teams will need people who can display a willingness to adapt and acquire new skill sets.

2.5.3 An investment in new technology, hardware and software is needed, and the potential cost of that investment is not justified by the potential savings that may be gained at this stage

In respect of this issue, what needs to be understood is that BIM cannot be bought out of a box. BIM is a process that is continuing to evolve. Therefore, although an investment in new technology, hardware and software will be needed, as with most other new innovations, this must be viewed as being an investment for the future, as there are anticipated significant cost savings and efficiencies for the organisation through using the BIM process.

In any event, and looking at the bigger picture, the reality of the situation is that organisations must adapt or they will fall behind in the market, and when that happens it can have a devastating effect on their future chance of success.

2.6 What will BIM do in the future?

What BIM will do in the future is the truly exciting. The UK Government's Construction Strategy published in 2011 requires that all central government construction procurement will require fully collaborative 3D BIM by 2016. This strategy was tested and justified by the UK Government's BIM hypothesis:

> 'Government as a client can derive significant improvements in cost, value and carbon performance through the use of open sharable asset information.'
>
> Government Construction Strategy (2011)

In considering its strategy, the UK Government understood that there was much more that could be done by BIM than was set out in its hypothesis but, at the time of forming its hypothesis, it saw any other functions as being either too complex for widespread adoption or not in direct support of the government's commercial or environmental agenda. However, the government did advocate that the supply chain in the construction industry should review the benefits of the BIM process and the continually emerging technologies, so that, if the companies involved thought that those emerging technologies or processes could add value to their businesses, they could adopt the applicable elements.

The above BIM hypothesis is, therefore, important as it enables members of the project team to demonstrate to other interested parties across a range of performance dimensions that useable benefits will be secured through the use of BIM in the future. Clearly, if information is digitised it is possible to do tasks much more efficiently and in a completely different way to how those tasks were undertaken previously, and this will drive forward future actions. If we consider, by way of an example, how people now search for properties to buy compared with the approach previously taken, this gives us a good idea how the advances of new technology can create a new way of operating.

Previously, people would visit a number of estate agents and drive around areas of interest to them to keep an eye out for 'For Sale' signs. This was a long and laborious exercise. Now, of course, there are many leading websites that allow a prospective purchaser to put in a set of criteria for their property requirements, and the properties that meet those requirements are sent to them electronically in seconds. Prospective purchasers no longer need to regularly repeat the initial exercise, as they can be alerted by email of changes and/or additions to the properties that are available immediately that this occurs.

It can be seen that, in the above case, technological advances were a major driver of the change in actions and procedures that followed. As BIM develops further and the use of BIM gains pace, a similar process will happen in the construction and civil engineering industries.

Therefore, if, for example, an architect wishes to specify a door set based on the requirement that it has a fire rating of 2 hours, then by simply entering that information through the BIM process, manufacturers' products that meet the required criteria will be notified to him from the databases held. Or, by way of another example, if a contractor is required to install a window that has a Kitemark licence for a particular British Standard, then a similar process will be followed by the contractor to see what product options are available.

When more than one product is available that satisfies the basic performance specification, it will then be possible to automatically compare other variables such as financial cost, environmental cost, expected lifespan, customer service and availability, among others, to decide upon which product should actually be selected to be used on the project.

As time goes on, the scope and diversity of software products and databases will broaden, and more functionality related to individual disciplines and specialisms will be provided. More and more applications will come to market, and these will support the professionals and the supply chains in terms of team inventory, off-site manufacture, fit first time and subassembly production.

Also, looking forward, sustainability and carbon footprinting will form a greater part of the essential requirements of the construction and civil engineering industries. Owners who are (or who will be required to be) more environmentally conscious will expect and/or require better predictability of energy performance and revenue costs, which in turn will rely on finding building products that have a better environmental performance. This is clearly an area where the use of the BIM process will particularly come into its own.

Therefore, a further significant driver is the use of environmental issues as a lever for progress, and the emphasis on 'green' building design and operation (also referred to as 'sustainable building and operation'). Through BIM, a large number of building energy simulation tools are, and will become, available that can be used to ensure the efficient and optimum use of and the reduction in waste of energy, water and other resources.

Furthermore, the use of BIM modelling will allow for the optimisation of lean construction techniques and principles, in that the process allows a participant to precisely plan and coordinate the design and execution of the project, leading to the potential for increased prefabrication, the minimisation of waste both in the shop and on site, the reduction of site clashes and collisions, and a general increase in productivity at all stages of the project.

In the future, BIM, in addition to 3D intelligent design (models), the fourth dimension, which adds time to the model, and the fifth dimension, which adds quantities and cost of materials (which already exist and which are dealt with later within this book), many other 'dimensions' will be added to the model (similar to 'apps' on smart phones) which may deal with the planning, design, construction and facility operation processes, among many other possible future dimensions.

2.7   To BIM or not to BIM?

Being realistic, most organisations, in particular those involved in the public sectors of the construction and civil engineering industries, will not have a choice in this matter because the UK Government's initiative, as outlined above, will advance the use of BIM greatly over the next few years, as will the need to increase efficiency and reduce costs in both the public and the private sectors. Therefore, any organisation involved in public sector works in the UK (even if only by way of a supply chain partner) will, in effect, be obliged to become involved in BIM, and because of this it is inevitable that the use of BIM will filter through into the private sector.

In addition to this, it appears to be quite clear that BIM is not a fad but is a process that will regularly be used more and more in the construction and civil engineering industries in the forthcoming years. The UK Government's initiative will clearly accelerate the adoption of BIM throughout the UK construction supply chain, and this will create critical mass and certainty of demand for the BIM process, which will provide the confidence that will enable businesses, training organisations and professional bodies to invest more rapidly in the development of the BIM process.

The reality is that BIM has the potential to unlock more efficient ways of collaborative working, and will offer better value to clients (both public sector and private sector). BIM, if and when successfully implemented, will help reduce waste and cost from the processes of providing an asset (with respect to the design, construction and life-cycle costs), and it is hoped (by the UK Government, at least) that this reduction could be in the region of 20%.

Naturally, there will be a cost implication for organisations becoming involved with BIM, but perhaps the more relevant consideration is the cost to an organisation if it does not become involved with BIM. If an organisation takes the latter course of action, it will lose ground to its competitors and it may not be able to secure some works that require a BIM involvement.

In reality, BIM systems are already commonly used by many designers, consultants, contractors and subcontractors involved in the construction process, and the acceleration of the development of information exchange standards and protocols must, and will, assist the adoption of effective ways of BIM working.

Therefore, there really is only one answer to the question 'To BIM or not to BIM?', and that is 'To BIM', as to do otherwise would mean being left trailing in the wake of others.

REFERENCES

BIM Task Force (2012a) *COBie UK 2012*. Available at: http://www.bimtaskgroup. org/cobie-uk-2012 (accessed 7 November 2012).

BIM Task Force (2012b) PAS 1192-2 Specification for information management for the capital/delivery phase of construction projects using Building Information Modelling. Available at: http://www.bimtaskgroup.org/pas-1192-22012 (accessed 7 November 2012).

Construction Project Information Committee (1997) *Uniclass: Unified Classification for the Construction Industry*. RIBA Publishing, London.

International Organisation for Standardisation (1994) ISO/TR 14177:1994 Classification of information in the construction industry. ISO, Geneva.

Government Construction Strategy (2011) https://www.gov.uk/government/ publications/government-construction-strategy (accessed 6 December 2013).

BIM in Principle and in Practice
ISBN 978-0-7277-5863-7

ICE Publishing: All rights reserved
http://dx.doi.org/10.1680/bimpp.58637.019

Chapter 3
How BIM works

3.1　An explanation of practical techniques

Original computer aided design (CAD) based technology is rarely used for BIM because it uses explicit coordinate-based geometry to create graphic entities. Documentation is then created by extracting coordinates from the CAD model to generate stand-alone 2D drawings.

As CAD software matured, graphical entities were combined to represent a design element, for example a wall or a hole. Following on from that, and depending on the software used, the CAD model was developed so that it became more 'intelligent' and was easier to edit. Surface and solid software modellers added more intelligence to the elements, and this enabled the creation of more complex forms of models.

But the result was still an explicit (coordinate based) geometric model, which was inherently difficult to edit and had a tenuous relationship to the drawings that were extracted from the CAD model. Those drawings obviously easily fell out of synch with the CAD model as it developed, and constant updates were necessary.

Editing of those unintelligent graphics (sometimes referred to as 'dumb graphics') was cumbersome and, because of its very nature, was often subject to errors. In contrast to the dumb graphics within a CAD model, BIM is based on a 3D virtual model. This is a geometric, object-oriented representation of the project which has incorporated within it or which has attached to it a number of BIM software applications that make the model 'intelligent' or interactive. The intelligent data embedded in the BIM model may include design criteria, detailed specifications or performance criteria. In addition, it may include commissioning, maintenance data and spare parts list, and other information that may be useful over the life cycle of the project.

As the construction industry began to associate data such as a key number or name with these graphic symbols, they were dubbed 'intelligent' or 'smart' data. In some cases these data, such as a height dimension for example, could affect the geometry of the symbol, making that datum a 'parameter' and the symbol a 'parametric'.

Following this, other basic relationships, such as 'hosting', were introduced between symbols, which would allow, for example, a window to remain attached to a wall if the wall was moved. If other parameters were built into the data (e.g. the top edge of the window would always be a certain dimension from the top of the wall and the bottom edge of the window would always be a certain dimension from the bottom of the wall) then the size of a window would automatically change if the height of a wall was changed.

However, the missing piece of the puzzle was the creation of a network of relationships among and between *all* the pieces of the building, so that if any part of the building was changed all the other affected parts of the building would also change automatically. This is the strength of a parametric building model. It records, presents and manages relationships between the parts of the building, no matter where they occur in the building.

An effective parametric building model manages object data at the component level, but more importantly allows information about relationships between all the components, views and annotations in the model. Therefore, a door to a stairwell can be locked in place at a specific distance from the riser of the stair to ensure egress clearances, or a door can be locked at a specific distance from a wall to ensure furnishing clearance or clearance for accessibility. In this way, the entire model contains information within it, not just the objects that form the model.

Therefore, in a parametric building model, simply selecting and moving a wall on a floor plan will cause all the related elements to adjust automatically. For example, the roof will move with the wall, preserving any overhang relationship, the other exterior walls will extend to remain connected to the moved wall, and so on. Similarly, doors and windows will 'know' their relationship to the walls in which they are placed and will adjust accordingly. This associativity is a defining feature of a true BIM model.

Accordingly, whereas a designer would normally detail a particular element, such as a window within an external wall, instead the designer establishes placement points within a wall which are determined by reference to the wall's length and height and the required characteristics of the window defined by the user. These characteristics may include surface area or height, for example, and such parameters and rules are the reference to which elements can then be created.

Therefore, taking the above mentioned window as an example, it may be created by reference to the wall in which it is located using attributes such as 'attached to', 'offset from' and 'protruding from', so that it is defined in relation to another object (e.g. a wall) by using measurements, rules and angles. Alternatively, the parameters of the window (e.g. the window area is always to be 30% of the total wall area) may determine the characteristics of the wall that surrounds it. This is a subtle but nevertheless notable distinction in the approach to design. Consequently, the data and parameters ascribed to an element of the building may provide a mechanism by which objects may vary with reference to the overall design, or, in the alternative, act as determinative characteristics around which the design must be formed. It is this dual and mutual linked relationship approach that is a key characteristic of parametric object modelling.

Normally, the data applicable to a BIM model come from product information templates. Therefore, taking a door set as an example, a spreadsheet can be produced where, for example:

■ the first column may show properties from the Industry Foundation Classes (IFC) standard (buildingSMART, 2013a), which cover the key performance attributes of the door set (e.g. acoustic rating, fire rating, security rating)
■ the second column may show properties that typically the owner of the building may be interested in (e.g. manufacturers' warranties, expected life of product, installation date)
■ the third column may show additional properties that have been added.

Manufacturers are producing product information templates to make them comprehensive for use in the UK market, and those manufacturers may look at those product templates and make their information available to the construction industry in a structured digital format. Small manufacturers can do this simply by putting their information in a spreadsheet format, and this could be seen as a first step in embracing BIM. On the other hand, manufacturers at the leading edge of the BIM process will deliver digital versions of their products as rich parametric objects. Of course, some objects will be parametric and others (e.g. bathroom fittings, door handles) will not. However, when looking at what BIM will mean for manufacturers in the future, it is conceivable that their BIM virtual objects will form the basis of their product catalogues of the future.

Consequently, architectural BIM software renders it unnecessary to create parametric objects from first principles (i.e. 'custom' parametric objects). Instead, to aid designers and save time, BIM software systems will provide a library of precedents of parametric objects that can be used or developed to suit the designer's requirements. These libraries of parametric object precedents will be provided within BIM applications and will generally relate to many of the standard components encountered within a building or an infrastructure project.

Naturally, this will be of considerable assistance to the designer, as the BIM system will include a pre-programmed definition as to what is, say, a floor, wall, ceiling, roof, etc., and how each is to interact with the other. From such basis users are then able to develop their own parametric objects, including 'custom' parametric objects in addition to the development of the 'off-the-shelf' precedent objects pre-provided within the BIM system. The BIM system developers will also provide tools by which the user can develop their own bespoke components to match particular requirements, while following the specification of the chosen software. These parametric objects will be assigned attributes that reflect the physical objects they represent, including, among other things, physical features, geometry, density, thermal capacity, cost and delivery time.

Naturally, there may be a varying and increasing number of databases that relate to the same element of a building, and therefore the application of BIM may very quickly become very complex. Thus considerable care must be given to the creation of the parametric object attributes, so that the BIM software notifies the user with a warning in the event of an overlap or clash, and identifies that overlap or clash to the designer, particularly following an automatic update of the design.

The care and attention to detail in the production of the parametric object attributes cannot, therefore, be overemphasised. Accordingly, even what appears to be the most mundane of elements must be carefully defined. This is a time-consuming exercise but is essential if an effective and efficient BIM model is to be produced.

BIM systems, therefore, may also be distinguished from other parametric modelling systems because they possess within them a pre-established set of parametric object classes, a family of parametric objects, with the potential for different attributes, as applicable, having been pre-programmed within them.

While BIM is a modelling system that is capable of producing drawings, it does this by way of the data added to the model, rather than simply by being an advanced drafting tool (i.e. by lines and shapes drawn and text written using CAD). Because of this ability to produce drawings, the BIM model also has the capability to directly create fabrication drawings, which avoids what can be a problematic and difficult step in the construction process. In a traditional work flow it would normally be necessary for fabricators to review the plans and specifications, prepare fabrication drawings, compare them to other fabrication and design drawings, have them reviewed by the design team, and eventually release the drawings for fabrication.

Naturally, errors can occur at any stage of this process. However, by using the data in the model, dimensional errors, conflicts and integration errors can be avoided or significantly reduced, because the coordination is through the BIM model. In addition, and as a reciprocal process, the BIM model can be updated with as-built information provided from site, which allows for the accurate fabrication of custom-made components, particularly those that rely on pre-existing site dimensions.

Consequently, accurate as-built drawings can be made available immediately at the close of construction through the use of BIM and the 3D model. This is because the 3D model, as it is updated throughout the project duration, actually represents in electronic format the physical design and construction of the project throughout all trades.

A parametric building model combines a design model (geometry and data) with a behavioural model (change management). The entire building model and complete set of design documents is held in an integrated database, where everything is parametric and everything is interconnected. This means that the design criteria or intent can be captured during the modelling process, and editing the model becomes much easier and preserves the original design intent.

Therefore, when components are sketched or placed, the software retains inter-element relationships. Subsequently, as one element is modified, the parametric change engine determines which other elements need to be updated and how to make the change. The approach is applicable to building applications because it does not start with the entire building model but usually starts with a few elements that are explicitly changed by the user and then continues with the selective propagation of changes to other elements, and then to the entire model where applicable.

Therefore, the essence of the architectural design of a building is in the relationships that can be embedded in the building model. The more relationships there are, the more parametric the model will become. The creation and manipulation of the relationships between various elements of the building is quite literally the act of designing. The use of parametric modelling gives designers direct access to these relationships, and this is a natural and intuitive way of thinking about buildings using a computer, in much the same way as a spreadsheet is a tool for thinking about numbers, or a word processor is a tool for thinking about words.

The analogy of a spreadsheet is often used to describe parametric building modelling. A change made anywhere in a spreadsheet is expected to update automatically all other calculations and figures everywhere else in the spreadsheet. The same is true for a parametric building model – a change in the data relating to one aspect of a building means that there will automatically be a real-time self-coordination of the information in every view and aspect of the model.

Just as no one expects to have to update a spreadsheet manually (and in fact would shy away from doing so for fear of introducing a calculation error), no one has to manually revise a document or schedule from a parametric building model. The BIM model generates the documents and drawings from the data that is inputted into the parametric building model.

This bi-directional associativity and immediate, comprehensive change propagation from the BIM model results in the high-quality, consistent, reliable model output that is key to BIM, an output that facilitates digital-based processes for design, analysis and documentation.

With state-of-the-art parametric building modelling, BIM software can coordinate a change made anywhere. It can produce 3D views and drawing sheets, schedules and elevations, sections and plans. If the model is updated in one place, all views, drawings and schedules are instantly synchronised and updated everywhere else.

From the above, it is clear that parametric building modelling is absolutely critical to the successful use of BIM.

3.2   The levels of BIM and BIM maturity

As should already be clear from the foregoing parts of this book, BIM is a rapidly developing field, and there is no real definition of what BIM is (or what it may become) and what it is not. However, and despite this, a convention has built up to categorise BIM under various 'levels', and the movement from one level of BIM to another is referred to as 'BIM maturity'. The levels of BIM are described below.

- *Level 0* relates to unmanaged CAD in 2D, with paper or electronic data exchange. Thus, this is not really BIM at all, and simply uses 2D CAD files for design and production information.
- *Level 1* represents the first step towards genuine BIM and the use of 3D data to present design. At this level the designer is usually operating in isolation, and thus this level of BIM is colloquially known as *lonely BIM*. At this level it may be that there are a number of designers, each of whom is dealing with his own designs in isolation from the others. Designers at this level usually use managed CAD in 2D or 3D format with a collaborative tool providing a common data environment, and normally comply with BS 1192:2007 (BSI, 2007), which is the code of practice for the collaborative production of architectural, engineering and construction information in the UK. Normally, some standard data structures and formats are used, and there may also be commercial data management provided by separate stand-alone finance and cost management packages which are not integrated in the general BIM model.
- *Level 2* is a managed 3D format held in separate BIM discipline software tools with data attached. A significant BIM Level 2 characteristic is the use of an 'as-built data drop' for the employer under a construction contract, the current method of which is presently called COBie UK 2012 (Construction Operations Building Information Exchange) (BIM Task Force, 2012a). This is dealt with in detail later in this book. The formalisation of this information exchange

necessitates the creation of information protocols, i.e. the establishment of agreed principles by which data will be shared and parties will cooperate as maturity nears Level 3. This level of BIM may introduce the first steps to utilising 4D construction sequencing data and/or 5D cost information.

■ *Level 3* will be a fully integrated and collaborative real-time project model that is likely to be facilitated by 'web services', given the demand on information technology (IT), and which will be compliant with emerging Industry Foundation Classes/buildingSMART Data Dictionary standards (buildingSMART, 2013b). Hurdles to software interoperability will have to be overcome, as will infrastructure difficulties and possible legal obstacles. This level of BIM will utilise 4D construction sequencing, 5D cost information, 6D project life-cycle information and other dimension (sometimes referred to as 'XD') management information, and will be driven by the development of standard libraries of object data, which will include manufacturers' information.

The majority of BIM usage at the moment is at Level 0 or Level 1, but the UK Government's BIM Strategy Paper published in 2011 (BIS, 2011) calls for the construction and civil engineering industries to achieve Level 2 BIM for UK Government projects by 2016.

The BIM maturity model describes by way of levels of maturity the ability of the construction supply chain to operate and exchange information. The said model is applied to an entire project situation. Therefore, while an organisation may claim to be operating at BIM Level 2, it may actually have a number of projects that are only able to operate at BIM Level 1. This would be perfectly normal, and indeed expected, as different organisations would mature on different timescales and at different levels.

The BIM maturity model is also used to define the supporting infrastructure required at each level of capability, and will be used to prioritise development of the BIM infrastructure. In addition, the BIM maturity model recognises that some supply chains will want to achieve greater levels of integration (up to and beyond Level 3) in line with the wider government construction strategy.

3.3 BIM models

The extended use of 3D intelligent design models has led to reference to terms such as 4D (adding time to the model) and 5D (adding quantities and cost of materials), and so on from there. Perhaps a simpler way is to think of the 3D model as a 'platform' onto which are built other applications that may be used through the planning, design, construction and facility operation processes. These further applications can deal with various types of information and data, and clearly the extent of the applications can be almost limitless.

Based on this, when coordinating construction sequencing by integrating schedule data with the model data and calling it '4D', or doing the same when using the model data to quantify materials and apply cost information and calling it '5D', seems very arbitrary, because these are just two of the many applications of how the basic 3D 'platform' can be used to improve all the processes.

Therefore, rather than continuing with this numbering after 5D to 6D, 7D etc., there is a growing trend to refer to all the extended applications using the 3D base platform as XD. However, at this time, the models will be considered under the headings of 3D, 4D, 5D etc.

The 3D Model provides for:

- *Model walkthroughs.* These provide a great visualisation tool enabling designers and contractors to work together to identify and resolve problems with the help of the model before walking on site.
- *Clash detection.* Traditionally, design drawings must be coordinated to ensure that different building systems do not clash and can actually be constructed in the allowed space. Accordingly, most clashes are identified when the contractor receives the design drawings and everyone is on site and working. With clashes being detected so late, delay is caused and decisions need to be made very quickly in order to provide a solution. BIM enables potential problems to be identified early in the design phase, and resolved before construction begins.
  The concept of 'design validation' is a task performed by the contractor, and is distinct from the 'design coordination' task performed by the design team. The contractor does not take responsibility for design coordination simply by engaging in design validation. The design validation is performed by detecting significant clashes.
  Normally, more detailed clash detection is performed by the contractor after the subcontractors have integrated their shop drawings into the consolidated model.
- *Project visualisation.* Simple schedule simulation can show the owner what the building will look like as construction progresses. This provides a very useful and successful marketing tool for all those involved in a project. Contractors can also use project visualisation to understand how the building will come together.
- *Virtual mock-up models.* On large projects, the owner will often request physical mock-up models so they can visualise, better understand and make decisions about the aesthetics and the functionality of part of the project. BIM modelling enables virtual mock-ups to be made and tested at a fraction of the cost of physical construction.
- *Prefabrication.* The level of construction information in a BIM model means that prefabrication can be utilised with greater assurance that prefabricated

components will fit once on site. As a result, more construction work can be performed offsite, cost efficiently, in controlled factory conditions, and then efficiently installed.

The 4D (time) model provides for:

- *Construction planning and management.* BIM models provide a means of verifying site logistics and operations by including tools to visually depict the space utilisation of the site throughout a the construction of a project. The model can include temporary components such as cranes, lorries and fencing. Traffic access routes for lorries, cranes, lifts and other large items can also be incorporated in the model as part of the logistics plan. Tools can further be used to enhance the planning and monitoring of health and safety precautions needed on site as the project progresses.
- *Schedule visualisation.* By watching the schedule visualisation, project members will be able to make sound decisions based on multiple sources of accurate real-time information. Within the BIM model, a chart can be used to show the critical path and visually show the dependency of some sequences on others. As the design is changed, advanced BIM models will be able automatically to identify those changes that will affect the critical path, and indicate what their corresponding impact will be on the overall delivery of the project.

The 5D (cost) model provides for:

- *Quantity take-offs.* To determine a project's construction cost and requirements, contractors traditionally perform material 'take-offs' manually, a process that is fraught with the potential for error. With BIM, the model includes information that allows a contractor accurately and rapidly to generate an array of essential estimating information, such as materials quantities and costs, size and area estimates, and productivity projections. As changes are made, estimating information automatically adjusts, allowing greater contractor productivity.
- *'Real-time' cost estimating.* In a BIM model, cost data can be added to each object, enabling the model automatically to calculate an approximate estimate of material costs. This provides a valuable tool for designers, enabling them to conduct value engineering.
- *Whole-life cost and life-cycle cost.* For many years, the terms 'whole-life cost' and 'life-cycle' cost were used interchangeably, and their meanings became confused. This unsatisfactory situation began to be addressed in 2008 with the publication of two documents on life-cycle costing: the international standard, BS ISO 15686-5:2008, 'Buildings and constructed assets' (BSI, 2008a); and the UK supplement to that standard, 'Standardized method of life cycle costing for

construction procurement' (BSI, 2008b), which set out clear definitions for the two terms.

- *Whole-life costing.* This is a methodology for the systematic economic consideration of all whole-life costs and benefits over a period of analysis, as defined in the agreed scope.
- *Life-cycle cost.* This is the cost of an asset, or its part, throughout its cycle life, while fulfilling the performance requirements. Broadly speaking, life-cycle costs are those costs associated directly with constructing and operating the building, while whole-life costs include other costs such as land, income from the building and support costs associated with the activity within the building.

The application of BIM modelling can assist in running through various scenarios to test the whole-life costing and the life-cycle cost implications of different approaches

The 6D (facilities management) model provides for:

- *Improved space management.* By understanding the details of how space is used, facility management professionals can reduce vacancy, and ultimately achieve major reductions in property expenses. The room and area information in BIM models are the foundation for good space management.
- *Streamlined maintenance.* The key challenge in developing a maintenance programme is entering the product and asset information required for preventive maintenance. The information about building equipment stored in BIM models can eliminate months of effort to accurately populate maintenance systems.
- *Efficient use of energy.* BIM can help facilitate the analysis and comparisons of various energy alternatives to help facility managers dramatically reduce environmental impacts and operating costs.
- *Economical renovations.* A 'living' BIM model provides an easier means of representing 3D aspects of the building. Better information about existing conditions reduces the cost and complexity of building renovation projects.
- *Life-cycle management.* Some building design professionals are embedding data on life expectancy and replacement costs in BIM models, thereby helping an owner understand the benefits of investing in materials and systems that may cost more initially but have a better payback over the life of the building.

The 7D ('green') model provides for:

- *Energy.* In current practice, many digital building models do not contain sufficient information for building performance analysis and evaluation. As with traditional physical models and drawings, evaluating building performance based on the graphical representations of conventional CAD or object CAD solutions requires a great deal of human intervention and interpretation, which renders the analyses too costly and/or time consuming. In a parametric building model, much of the data needed for supporting design analysis are captured naturally as the design of the project proceeds. The model contains the necessary level of detail and reliability to complete these analyses earlier in the design cycle, and makes possible routine analysis done directly by designers for their own baseline energy analysis, thus providing immediate feedback on design alternatives early on in the design process.

■ *Life-cycle analysis.* Life-cycle analysis is an assessment of the environmental impact of a product or service throughout its life cycle, from cradle to grave. The Building Research Establishment's (BRE) *Green Guide to Specification* (Anderson *et al.*, 2009) is a database of the life-cycle analysis of a variety of construction products. The *Green Guide* rates each product on an A+ to E ranking system, where A+ represents the best environmental performance/least environmental impact, and E the worst environmental performance/most environmental impact. The environmental rankings are based on life-cycle assessments using BRE's Environmental Profiles Methodology (BRE, 2008).

The BIM process can assist in this process, by concentrating on certain elements of the building (e.g. windows, frame, roof, floor slab, internal walls, external walls), and considering the choice of materials/products that may have a direct or indirect impact on the environment, as well as the capital and operational costs. The materials chosen should be those that may generate energy, waste or water savings. Running the scenarios and analysis through the BIM process enables project teams to review the environmental and economic impacts of their decisions, and assists with ensuring that the 'best' materials and products have been chosen for the project.

3.4   Hardware and software

BIM is an advanced computerised system, and so clearly it relies on good hardware and efficient and effective software. The hardware aspect is relatively inexpensive, so it is not worth skimping on this, as it is important that the hardware used will allow the chosen software to operate at its full capacity. In respect of software there are some large software houses that are used as a standard, but smaller software providers are also entering into the BIM market.

When considering what software to use then the following factors should be borne in mind:

■ Simplicity – make sure the software is easy to learn and use.
■ Functionality – ensure that the tool meets your specific needs and usage by reading about the tool before you start using it.

- Interoperability/collaborative – the tools you use should work well with other software, as being able to interchange document formats or convert documents is absolutely essential.
- Provides longevity – due to the rapidly changing technology environment, make sure you are confident that the vendor will be around for the long run.
- Support/training – the tool should have quick, effective help, and the provider should include appropriate training (electronic and in person).
- Environment – ensure that the tool will work in your environment with your hardware, communications and collaborative partners. You may want to consider using one of the various technology hosting services to provide the environment for your tools (especially in the beginning, until you determine your specific needs).

To date, BIM technology has been developed to facilitate specific processes and activities related to a project. At the core, BIM software is a database. Its application to a process requires that the database initially be populated, and then maintained as the project progresses. The amount of redundant effort required to develop and maintain the various databases of the many subcontractors that employ BIM technology represents the greatest source of waste and error associated with BIM implementation. In order to facilitate the full integration of BIM technology, software vendors need to develop ways for the various members of the project team to input and maintain the data relating to the specific aspects of the project within their responsibility. In short, interoperability is essential, and must be accommodated by the software industry.

Software providers must also understand the process of design and construction, and fit their software to these workflows. BIM software must be capable of modification as design progresses, so that the increasing levels of detail characteristic of the various design stages – from schematic design to construction documents – can be included in the BIM model at the appropriate point in the design process. The software must also be able to accommodate changes.

In most software, a model is a single database file represented in the various ways that are useful for design work. Such representations can be plans, sections, elevations, legends and schedules. Because changes to each representation of the database model are made to one central model, changes made in one representation of the model (e.g. a plan) are propagated to other representations of the model (e.g. elevations). Thus, drawings and schedules are always fully coordinated in terms of the building objects shown in drawings.

When a project is shared between several users, a central file is created that contains the master copy of the project database on a file server on the office's local area network.

Each user works on a copy of the central file (normally then referred to as a 'local file'), stored on the user's workstation. Permissions on objects are often managed by locking them in the central file, ensuring that only one user has rights to them at a time. Users can periodically synchronise their changes back to the central file and receive changes from other users.

Multiple disciplines working together on the same project make their own project databases and link into the other consultants' databases for verification. The software automatically carries out collision checking, which detects if different components of the building are occupying the same physical space. When set up correctly, schedules can also provide information to verify the functional aspects of a building (e.g. the level of occupancy for the room area, as well as electric and ventilation loads).

Software often supports the open XML-based IFC standard, developed by the buildingSMART organisation. This file type makes it possible for an employer or a contractor to require BIM-based workflow from the different discipline consultants of a building project. Because IFC is a non-proprietary and human readable format, it is achievable and compatible with other databases, such as facility management software.

3.5   PAS 1192-2:2012

PAS stands for Publicly Available Specification. In the context of BIM, the draft British Standard PAS 1192-2:2012 (BIM Task Force, 2012b) relates to the information management for the capital/delivery phase of construction projects using BIM. The purpose of PAS 1192-2:2012 is to support the UK Government's stated objective of achieving BIM maturity Level 2 (together with the desired reduction in capital expenditure out-turn cost) on all public sector asset procurement by 2016 by specifying requirements for achieving that level, by setting set out the framework for collaborative working on BIM-enabled projects, and by providing specific guidance for the information management requirements associated with projects delivered using BIM.

PAS 1192-2:2012 builds on the existing code of practice, BS 1192:2007 (BIM Task Force, 2012a), for the collaborative production of architectural, engineering and construction information, as defined within BS 1192:2007.

BS 1192:2007 provides details of the standards and processes that should be adopted to enable consistent, structured, efficient and accurate information exchange specific to BIM. PAS 1192-2:2012 focuses specifically on the 'delivery' phase of projects (from strategic identification of need through to handover of asset), where the majority of graphical data, non-graphical data and documents are accumulated from design and construction activities.

PAS 1192-2:2012 provides a framework, from which a number of supplementary documents will provide detailed guidance. The requirements of the PAS will need to be applied specifically to individual projects, commencing at the point of assessment (for existing assets) or statement of need (for new assets), and progressively working through the various stages of the information delivery cycle. The requirements within this PAS culminate in the delivery of the as-constructed asset information model, which is handed over to the employer by the supplier.

PAS 1192-2:2012 also describes the shared use of individually authored models in a common data environment (CDE), this being a single source of information for any given project that is used to collect, manage and disseminate all relevant approved project documents for multi-disciplinary teams.

Whereas BS 1192:2007 provides details of the standards and processes that should be adopted to enable consistent, structured, efficient and accurate information exchange, PAS 1192-2:2012 describes only structured data and information exchanges specific to BIM. In addition, PAS 1192-2:2012 focuses specifically on the 'delivery' phase of projects (from strategic identification of need through to handover of asset), where the majority of graphical data, non-graphical data and documents are accumulated from design and construction activities.

A forthcoming document, PAS 1192-3, which has yet to be developed, will offer guidance on the use and maintenance of the asset information model (AIM) to support the planned preventive maintenance programme and the portfolio management activity for the life of the asset.

3.6   Construction Operations Building Information Exchange (COBie)

COBie (Construction Operations Building Information Exchange) is a means of sharing, predominantly non-graphical, data about a facility. It was developed by the US Army Corps of Engineers to manage the data coming from BIM models into the client organisation, particularly for the handover of operations and maintenance information.

This facility has been developed in the UK (as COBie UK 2012 (BIM Task Force, 2012a)) to provide data reporting at specific stages of the project. These reporting stages are normally referred to as 'COBie data drops'.

The data are exchanged using spreadsheets to keep the complexity of systems and training to a minimum. COBie is an interim step to a technology that will allow open exchange of all project data. Such standards have been in development for some time and are starting to emerge but need a little more time to mature to a stage for widespread end-to-end adoption. The use of a spreadsheet is only coincidental, as the information will mostly used be either in a BIM or FIM tool. However, using spreadsheets does mean that anyone will be able to make use of the information with minimal cost.

The focus of COBie is on delivering building information, not geometric modelling. COBie is a subset of a building model referred to as a 'model view' or 'data view'. COBie data is delivered along with existing contract deliverables depending on the specific contract conditions. COBie does not change the content of existing contract deliverables. It does, however, change the format of the information that is delivered.

Implementation of COBie relies on three primary functions:

- *gathering* asset information
- *maintaining* the information
- *using* the information.

The timing of the data drops will vary depending on the requirements of individual clients, to suit their internal processes and approvals. However, typically, data drops may be required as:

- *Data Drop 1.* The first data drop represents the requirements and constraints, i.e. to check the design brief against the client's brief and to provide cost plan and risk management. The data available at Data Drop 1 are broadly consistent with those expected at the existing Royal Institute of British Architects (RIBA) Stage B – Feasibility (i.e. the brief). The rationale for this data drop is to approve the 'outline business case' – at this time, checks will be made to ensure that the emergent design and specifications are consistent with the client's brief in terms of function, cost and carbon performance. The data contain all client requirements and constraints information.
- *Data Drop 2.* The second data drop is broadly consistent with the data expected at the existing RIBA Stage D, i.e. to check tender design against project brief, cost plan, environmental requirements and provide tender transparency. The rationale for Data Drop 2 is to select the main contractor. Checks are made at this stage to ensure that the interpreted design and specifications are consistent with the client's brief in terms of function, cost and carbon performance, and that the potential suppliers and supply chain can demonstrate capability and integrity through the competitive process and be selected to deliver the asset. The process (assuming a quantitative process) will include costs and carbon at a level to be agreed in the employer's information requirements. At this stage, the model would first be delivered by the client-side technical team and the model returned by the contracting supply chain.

- *Data Drop 3*. The data available at Data Drop 3 is broadly consistent with those expected at existing RIBA Stage F, i.e. to check detailed design and contract packaging for scope, cost and carbon footprint. The rationale for this data drop is to approve the 'agreed maximum price' or 'works order'. Checks are made to ensure that the developed design and specifications are consistent with the client brief in terms of function, cost and carbon performance.
- *Data Drop 4*. The data available at Data Drop 4 are broadly consistent with those expected at existing RIBA Stage K, i.e. to provide and check handover data, actual cost, actual programme and actual carbon performance. The rationale for this data drop is to take possession of the 'operations and management' information. The data being collected are the operational and detailed functional information supplied by the product manufacturers.

In the comments above, reference is made to the 'existing' RIBA stages. This is because the RIBA plans to abolish its 50-year-old traditional scheme covering Plan of Work Stages A to L and replace it with an eight-point schedule in a bid to integrate practices across the industry and accommodate modern collaboration and BIM. The 'new' eight stages are:

- *Stage 0: Strategic definitions*. In this stage, a project is strategically appraised and defined before a detailed brief is created.
  This relates to the 'Strategy' part of the digital plan of work.
- *Stage 1: Preparation and brief*. This stage merges the residual tasks from the former RIBA Stage A (Approval) with the former RIBA Stage B (Design brief), tasks that relate to carrying out preparation activities and briefing in tandem.
  This relates to the 'Brief' part of the digital plan of work.
- *Stage 2: Concept design*. This stage replicates exactly the former RIBA Stage C (Concept).
  This relates to the 'Concept' part of the digital plan of work.
- *Stage 3: Developed design*. This stage is essentially the same as the former RIBA Stage D (Design development), with the addition that the developed design will be coordinated and aligned with cost information by the end of the stage.
  This relates to the 'Definition' part of the digital plan of work.
- *Stage 4: Technical designs*. This stage comprises the residual technical work of the core design team members.
  This relates to the 'Design' part of the digital plan of work.
- *Stage 5: Construction/specialist design*. This stage recognises the importance of design work undertaken by specialist subcontractors and/or suppliers employed by the contractor.
  This relates to the 'Build and commission' part of the digital plan of work.

■ *Stage 6: Handover and closeout.* This stage replicates the former RIBA Stage J (Mobilisation) and Stage K (Construction to practical completion).
This relates to the 'Handover and closeout' part of the digital plan of work.

■ *Stage 7: In use.* This stage replicates the former RIBA Stage L (Post practical completion), with the addition of further duties arising from post-completion and post-occupancy evaluation activities.
This relates to the 'Operation and end of life' part of the digital plan of work.

Similarly to the RIBA schedule, the new Construction Industry Council (CIC) digital plan of work has eight clear project stages (0 to 7) that has been developed by all the construction industry institutions working in collaboration. The CIC's digital plan of work (unlike the RIBA's stage of works document) includes the responsibilities of all consultants in the project team, not just the architect.

The amount of the COBie UK 2012 worksheets that needs to be filled in depends on the project stage. Project team members only enter the data for which they are responsible: designers provide spaces and equipment locations; contractors provide manufacturer information and installed product data; and commissioning agents provide warranties, parts, maintenance information.

All products and equipment listed in design schedules should be found in the COBie UK 2012 file under the Type and Component worksheets. Type worksheets identify the category of product. Components are specific instances of each of the Types, typically found in one room or area. Components must be listed by room or area. Components that link or span rooms must be listed for each applicable room (e.g. interior doors should be listed in both the spaces that the doors connect).

Typically, the worksheets may relate to:

■ *The early design stage.* As the design begins, the vertical and horizontal spaces that are necessary to fulfil the client's requirements for the building, facility or infrastructure project are defined. Within these buildings, facilities or projects, the different types of system that are needed to satisfy the owner's requirements are also defined.
As a significant benefit can be achieved for asset managers, COBie allows the exchange of space function and area calculations provided directly by the designers' CAD or BIM software.
Early in the design, projects are developed by listing spaces and identifying specific functions required to meet the owner's requirements. To allow these spaces to perform as intended, specific building systems are also required for all projects. For buildings, these systems include: electrical, heating, ventilating and

air-conditioning (HVAC), potable water, waste water, fire protection, intrusion detection and alarms, and other systems. In COBie, there must be at least one system for each facility.

- *The construction documents design stage.* As the design progresses, the material, products and equipment needed for the building are specified. The types of product are most often displayed as finish, product and equipment schedules. The use of these schedules for any of a variety of reasons, including quantity take-off, asset management and, of course, facility maintenance and operations, requires multiple, error-prone manual transcriptions. The types of equipment are listed along with the specific location of each of these types of equipment. The properties of products are listed as COBie common attributes. With these data structures, COBie transfers schedule information from designers to builders, and later to operators. Information within the COBie file allows the designer to identify fixed or movable property.

Components are organised into systems that are also listed in COBie. These systems provide specific building services to building occupants such as alarms, electrical, fire protection, HVAC and plumbing systems. Currently, an optional COBie set of data is the connections between equipment. Connections allow the designers to specify how specific pieces of equipment are logically connected. This would allow, for example, a worker to know what other equipment would be affected if a valve closed.

During the design there may be documents of interest pertaining to specific parts of the building. These documents can be linked by reference to the COBie 'documents' data. Designers may also specify the requirements for documents in COBie. One of the most common lists of required documents is the submittal register. The submittal register is a key aspect of COBie because it is the approved submittals during construction that comprise the bulk of construction handover data sets.

- *The contractor's quality control stage.* As the project progresses from design to construction, the next stage of the project that contains COBie data occurs when the contractor provides submittals for the designer-specified required documents. COBie information exchange allows electronic copies of acknowledged or approved submittals to be linked directly to specific types of material, product, equipment and system within the building.

The majority of these linked documents are provided as PDF files from documents already created by product manufacturers. Shop drawings should be lined in their native CAD/BIM formats as well as in PDF views. Scanned or photographic images are required for submittals that require physical samples. When the COBie data are transmitted, these files are provided with the COBie file on a single COBie data disk.

It is during this stage that construction contractors have a choice about how to implement COBie. They could continue to create facility handover data at the end of the construction process, and simply scan and link COBie data. While meeting the COBie requirement, the effort does little to streamline the process or reduce the cost of the submittal process. If, however, the construction contractor and the owner utilise an electronic submittal register, there will be virtually no cost for the collection of submittals at the end of the project, as these documents will all be provided in the submittal register software. Enquiries to large and small general contractors across the country have indicated that they are ready to provide electronic submittals. However, construction managers and owners need to be able to accept such submittals. Unfortunately, it is the case that the ability of contractors to meet the requirements of COBie exceeds the current level of expertise of owners to accept and process electronic submittals. Given the cost- and time-saving opportunities for contractors, it is possible that the use of e-submittal programmes can significantly speed up the delivery of accurate COBie data.

■ *The product installation stage.* Once the construction contractor has procured the specified materials, products and equipment, they are installed. The manufacturer and model for all products are listed under Type data. The documentation regarding the manufacturer and model can be documented either at installation or during the prior submittal process. The serial numbers for as-installed equipment and/or tags are documented in the Component data. As room names change, contractors can also provide the room number tag if that number differs from the room listed on the original design.

While large projects will be able to support the purchase and use of commercial software to document installed equipment, the majority of construction in the UK is accomplished by small contractors who may not have access to such software. For these companies, direct use of a 'locked down' version of the COBie spreadsheet should be satisfactory. Unlike the manual creation of equipment lists required today, contractors need only to change the room location for equipment if there is a change order related to that equipment.

It is important to note that the requirement of contractors to provide equipment and valve tag lists is already a requirement in virtually all construction contracts. COBie requires nothing new, simply a change of format in existing contract requirements. The contractor is free to use COBie as part of his traditional process or to take the COBie challenge to eliminate the end of project 'job crawl' in lieu of simply typing in the serial numbers of equipment and tags as they are installed.

Warranty information associated with bulk items (such as carpet) or one-of-a-kind products (such as medical equipment) are identified by Type. The component's installation date provides more detailed warranty start dates if needed.

■ *The system commissioning stage.* Once the equipment has been installed and tested, the systems are turned on and made operational for operations and maintenance staff. In COBie, there are several documents that describe system operations. These documents include Instructions, Tests and Certifications. As with all other submittals, COBie documents are provided in native or PDF format and referenced in the COBie Documents data set.

The final stage of commissioning is to develop scheduled or preventive maintenance and other types of plan that support long-term facility operations. In COBie there is space for the following the types of plan: Preventive Maintenance, Safety Plans, Trouble Shooting Plans, Start-Up Procedures, Shut-Down Procedures and Emergency Plans. These plans, or the documents containing these plans, are provided through the COBie job data.

In addition to listing specific types of job plans, COBie requires the identification of critical resources needed for these job plans. Often special materials, tools or training will be needed before starting a particular job. These resources are identified in COBie as Resource data. As with other information provided by the manufacturer provided and augmented by the commissioning agent, spare parts data may be provided as a document as individual spare part records.

Manufacturers often provide job plans and parts diagrams with product data sheets and catalogues. The need of construction contractors to reference a consolidated set of manufacturer data in such large documents should be a short-term effort. Once the COBie format is established in the construction industry, manufacturers will begin to provide COBie data directly to construction contractors, along with PDF catalogue extracts, which will also include manufacturer's suggested maintenance plans, as well as standard warranty and replacement parts.

The COBie UK 2012 specification is a performance specification. This means that it does not matter what software is used to create COBie UK 2012 information, as long as the format of the information meets the COBie UK 2012 specification and the content of the COBie UK 2012 file reflects any particular project.

Software vendors have begun to export directly to COBie UK 2012. However, on small projects COBie UK 2012 may also be created or updated by hand directly in the spreadsheet version of the COBie UK 2012 data.

The submittal register is held within the Document Worksheet. The set of all documents identified in the designer COBie UK 2012 file as 'Required' can be used to create a submittal register. Use of a web-based submittal register to automatically link manufacturer documents to building information (models) is essential to the cost-effective implementation of COBie UK 2012.

Each COBie UK 2012 file should contain a single building/asset. If there are multiple buildings/assets in a given project, the data for each building should be held in their own COBie UK 2012 worksheet.

COBie data should be extracted from each model separately and mapped into a single file. A number of tools and workflows for COBie extraction are already available, and more will soon be forthcoming as the individual software vendors develop their workflows and toolsets for aggregating COBie data.

Handover can be improved by delivery of consistent data through COBie, and several Computerised Maintenance Management Systems (CMMS) have implemented direct import of COBie information.

Today, most projects require the handover of paper documents containing equipment lists, product data sheets, warranties, spare part lists, preventive maintenance schedules and other information. This information is essential to support the operation, maintenance and management of the facilities assets by the owner and/or property manager. Gathering this information at the end of the job, which is today's standard practice, is expensive, as most of the information has to be recreated from information created earlier. COBie simplifies the work required to capture and record project handover data.

The COBie approach is to enter the data as it is created during design, construction and commissioning. The UK Government has stated that its aim is to mandate the 'Soft Landings' handover protocol across all central government projects by 2016 (Cabinet Office, 2013). Therefore, this protocol will form part of the public sector procurement process from 2016. The Government Soft Landings (GSL) policy will also draw on the Soft Landings framework, published by BSRIA and the Usable Building Trust in 2009, as a guide to how project teams should engage after practical completion – hand-holding the client during the first months of operation in order to fine tune the building. Whereas adopting a new asset should be a positive experience, often the gap between the client's design expectations and the delivered performance varies significantly. It is clear that the planning for facilities management does not start early enough, leading to wasted time and resources before the asset becomes fully operational. The transition from completion to operation often takes considerable time, effort and resources before the predicted performance is achieved.

COBie helps facilities managers identify their requirements at briefing, design, construction, completion and aftercare stages, and the GSL protocol helps achieve a smooth handover and commissioning process. Together, they will form an integrated system for asset management.

Facilities managers should engage with end users throughout the design and delivery process. They should set clear targets and measures for:

- functionality and effectiveness, so that the working environment is conducive to productivity and the well-being of the end users
- operational and capital costs, to reduce costs in construction and operation
- environmental performance, to meet carbon and other sustainability targets
- assessing performance, for at least three years after completion to establish outcomes and lessons learnt
- involving the design team in the early operating phase to fine tune performance and to ensure target outcomes.

The Soft Landings approach, provides a client with the opportunity to achieve assets that:

- achieve the performance outcomes specified at the outset of the project
- address operational needs and optimise running costs, adding to the efficiency of the business
- are cheaper to construct and operate and do not require changes after handover, because user and operator needs have been assessed throughout the design process
- meet the optimum performance much quicker because the designers and the construction team are involved in optimising their operation.

REFERENCES

Anderson J, Shiers D and Sinclair M (2009) *The Green Guide to Specification*, 4th edn. Wiley-Blackwell, London.

BIM Task Force (2012a) *COBie UK 2012*. Available at: http://www.bimtaskgroup. org/cobie-uk-2012 (accessed 7 November 2012).

BIM Task Force (2012b) PAS 1192-2 Specification for information management for the capital/delivery phase of construction projects using Building Information Modelling. Available at: http://www.bimtaskgroup.org/pas-1192-22012 (accessed 7 November 2012).

BIS (Department for Business Innovation and Skills) (2011) A Report for the Government Construction Client Group: Building Information Modelling (BIM) Working Party. Strategy Paper. Available at: http://www.bimtaskgroup.org/wp-content/uploads/2012/03/BIS-BIM-strategy-Report.pdf (accessed 7 November 2013).

BRE (Building Research Establishment) (2008) *Environmental Profiles Methodology*. Available at: http://www.bre.co.uk/greenguide/page.jsp?id = 2106 (accessed 7 November 2013).

BSI (British Standards Institution) (2007) BS 1192:2007 Collaborative production of architectural, engineering and construction information. Code of practice. BSI, London.

BSI (2008a) BS ISO 15686-5:2008 Buildings and constructed assets. Service life planning. Life cycle costing. BSI, London.

BSI (2008b) PD 156865:2008 Standardized method of life cycle costing for construction procurement. A supplement to BS ISO 15686-5. Buildings and constructed assets. Service life planning. Life cycle costing. BSI, London.

buildingSMART (2013a) buildingSMART Data Dictionary. Available at: http://www.buildingsmart.org/standards/ifd/dictionary-international-framework-for-dictionaries-ifd (accessed 7 November 2012).

buildingSMART (2013b) *Industry Foundation Classes (IFC) Data Model.* Available at: http://www.buildingsmart.org/standards/ifc (accessed 7 November 2012).

Cabinet Office (2013) Government Soft Landings. Executive Summary. Available at: http://www.bimtaskgroup.org/wp-content/uploads/2013/05/Government-Soft-Landings-Executive-Summary.pdf (accessed 7 November 2012).

BIM in Principle and in Practice
ISBN 978-0-7277-5863-7

ICE Publishing: All rights reserved
http://dx.doi.org/10.1680/bimpp.58637.043

Chapter 4
Incorporating BIM

4.1 Getting started with BIM

As previously noted, the building information modelling (BIM) process, when fully applied, is intended to cover a project from inception to demolition or disposal. However, it is not necessary to model the entire project to use BIM on a project. It is already the case that many projects are being run using models and/or intelligent models (often in part), even though the project is not subject to the BIM process in full.

It may be that a designer and/or some of the specialist subcontractors and/or some of the suppliers are using 3D models for their own benefit but not be sharing the information obtained from the models with other members of the project team. There are many such models in use, and they may deal with matters such as:

■ architectural design
■ structural design
■ mechanical and electrical services design
■ manufacturing drawings
■ working drawings.

It is not necessary for a project to be subject to the full BIM process to gain some of the benefits of BIM, and it is therefore sensible that BIM is used even if only for part of the project, for example for the design of the structural steel or the mechanical services systems, so that some benefits may be gained and the participants begin to get used to the BIM process in more general terms.

Also, and in some ways ahead of the industry generally, contractors are making use of intelligent/parametric models for portions of the project scope to assist them with many of their traditional activities, including:

■ demonstrating project approaches during bidding and/or marketing presentations
■ determining the scope of works for tendering purposes
■ ascertaining the scope of works for purchasing purposes

- reviewing portions of the scope of works to assist with value engineering exercises
- coordination of the sequencing of trades (even if only on a very limited basis).

The list of partial uses of BIM seems almost infinite. For contractors already using BIM, the list seems to grow daily. For those getting started, the following list represents some of the more common 'early' uses that most contractors experience in their experimentation with BIM:

- visualisation
- scope clarification
- partial trade coordination
- collision detection/avoidance
- design validation
- planning, construction sequencing phasing plans and logistics
- presentations and marketing
- analysis.

In each of the above cases only portions of the scope and only specific trades may be modelled, but benefits are still obtained.

In reality, it is through the partial use of BIM, as outlined above, rather that the application of a full BIM process that may project participants are likely to start engaging with BIM. These partial uses can, in many ways, be much less overwhelming to create and use, and the benefit of having them is much more immediately tangible for participants. Therefore, by applying BIM in smaller bites may be an easier way for many participants to become involved in the BIM process.

4.2   The integration of BIM

The greatest benefit of BIM will be the ability of the BIM system to offer full integration of the data generated by all actors within the model.

The successful integration of BIM may be viewed as a transition:

- from being simply viewed as 'a tool', whereby design applications are used as a tool to carry out certain tasks
- to being seen as a 'platform' by which the system can be used to manage the data input from different users
- to being seen as a much more sophisticated 'environment', in which the system supports multiple platforms, and processes and assimilates information from different models, and has the ability to provide multiple representations of the environment.

4.2.1 BIM as a tool

In summary, in terms of being a tool, the various BIM systems available may be differentiated in terms of:

■ The intuitiveness of the user interface – the more intuitive the system is, then less time will be spent learning, the fewer the mistakes that will be made and the more effective it should be.

■ Drawing functionality – there is a direct and equal relationship between the degree of success that changes made within the model can be demonstrated within drawings, with successful on-site construction implementation, and therefore commercial success.

■ Custom parametric object model tool development – the ease with which the user can create custom-made objects within the BIM and fit the same within different groups of objects, and link its attributes in the same manner as the system's pre-existing precedent models.

■ Higher level modelling tools – namely the capacity of the application to:
  – support full parametric modelling, including sophisticated shapes and objects
  – provide take-offs, track revisions, incorporate standard industry specifications and, perhaps significantly, warn of conflicts and clashes between objects and shapes.

Compared with computer aided design (CAD) tools, BIM design tools are more complex, offer greater functionality and continue to become progressively more intuitive as software developers respond to the commercial opportunities presented by growing demand. The consistency with which the software achieves this, however, continues to improve across the various providers.

Whereas CAD is the representation of lines, shapes and text, with fixed geometry and properties, on a computer screen, BIM is not fixed. Rather, BIM is a representation of the data inserted; it consists of defined rules, relationships, geometry and parameters, called 'attributes', which are used to create lines and shapes, which in turn generate parametric objects, the representation of which are the lines and shapes with which we are familiar. It is as a consequence of this that one parametric object can be related to another, so that if one is altered so too is the other (i.e. the basis of automatic updating). The way in which one parametric object automatically updates has been called 'behaviour'. However, there is no 'free will' in this process; the updates are automatic and are governed by relationships with other objects. For this reason, it is likely that it will be necessary to create or utilise varying classes of the same element, so that different attributes may be provided for each element within a class.

Architectural BIM software renders it unnecessary to create parametric objects from first principles (i.e. 'custom' parametric objects). Instead, to aid the designers and save time,

BIM systems provide precedent parametric objects that can be used or developed to suit the designer's requirements. This is of considerable assistance to the designer because the BIM system will include a pre-programmed definition of what is, say, a floor, wall, ceiling, roof and so on, and how each is to interact with the other. From such a basis users are able to develop their own parametric objects, including 'custom' parametric objects in addition to developing the 'off-the-shelf' precedent objects pre-provided within the BIM system. In this way libraries of objects are created and become available to users.

4.2.2 BIM as a platform

Progressing along the spectrum of BIM applications from tools, we find that higher performing systems may also be assessed in terms of their ability to function as a platform. The requisites of such systems have been developed accordingly and, significantly, in parallel with the growing appreciation of the potential that such applications have to offer.

The platform can be used in many ways, including design and drawings, design and calculations, design and fabrication, coordination management and facility management, each with tools appropriate to the task at hand.

In terms of effectively performing as a platform, however, the degree to which a system will succeed will depend on the capability of applications to facilitate

- Capacity – the ability of the system to quickly process, without malfunction, large amounts of data generated by significant levels of detail, and in particular data from sources originating from outside the application from other participants, i.e. to act as a platform without difficulty. The volume of data that is required to be processed in the context of a platform is particularly fertile ground for the effective deployment of cloud computing based technologies.
- Tool interface – the ability of the platform to successfully exchange data with other applications is key to a successful platform, and the system must be able to accommodate the assimilation of other geometries, attributes and parametric objects developed within applications from outside the platform. This is important in order to automatically identify clashes and conflicts, and to enable the system to incorporate, within linked data production, the impact of such data as structural loadings, costs and other forms of statistical intelligence commonly necessary for the effective design, construction and management of buildings.
- The development of standard parametric object precedents – including the importation of redefined objects from outside the application, renders it unnecessary for the user to define parametric objects, which can be complex.
- Translation between different applications – the extent to which the platform supports the unhindered import/export of data from and to outside sources.

Alternative data sources demand increasingly complex translation mechanisms, and so it is beneficial if these can be customised by the user to suit his specific needs.

- Increase collaboration – the strength of BIM and the fuel that primarily drives the industry's interest in it could be argued to be the capacity of the application to increase collaboration between the many stakeholders, both within the industry and, in particular, at the project level. Accordingly, it is important that systems support efforts to perpetuate the industry's current government-sponsored drive towards ever increasing collaboration. In this regard, it is essential that the platform not only permits but also supports multiple users and data entry.

The basic BIM design tool originated from the CAD tools, and it is through this development that an appreciation of its application to building modelling was recognised. However, while BIM systems facilitate tasks to be undertaken, as strength as a tool, with varying degrees of sophistication across the products available, is nevertheless as a multi-user data management platform, with its intuitive operation, its ability to process and store large amounts of information, and its ability to interface with further data sources, and this is where the future lies. The demands on BIM systems as a platform have continued to crystallise in parallel with the construction and civil engineering industries' growing appreciation of its functional applicability.

There is a variety of BIM design types with different parametric object precedents to be applied to a spectrum of building systems. Compared with manufactured objects, the rules applicable to, generally speaking, relatively simple building systems are much more predictable, and are therefore easier to define. Regardless of this, even higher value computer workstations can struggle with the amount of data necessary to process just the construction information generated within a medium-sized construction project, the very type of project that forms the majority of construction contracts in the UK.

Notwithstanding the hurdle that the sheer volume of data may generate, compared with the situation in general manufacturing, modelling using a BIM system does benefit from the large amount of rules, regulations and standard practices that have been long established across the broad range of disciplines within the built environment that specialise in the construction and facilities management of property. These can be encompassed within the parametric object attributes.

The construction industry also continues to be constrained by certain conventions, particularly with regard to the communication of design in the form of the need for drawings, a convention by which general manufacturing has not been bound by for a comparatively long time. It is this convention that has continued to limit the development of parametric modelling tools in the built environment, as they do not generally

47

support the production of drawings within architectural norms. As a consequence, and until more recently, only comparatively few parametric model systems had been developed for BIM.

At one level there is simple parametric solid modelling, in which a few attributes define shapes, and editing requires the attributes to be altered. In this case there is no automatic updating until the user reworks the design piece. Progression from such base levels includes those systems that facilitate automatic updating whenever attributes are altered, with each change being made across the sequence of the design. At the upper level, however, the rules and parameters of one shape may be linked in various ways to other shapes, and in so doing the system then determines the sequence in which to make updates as a consequence of changes made. This is usually defined as full BIM or full parametric object modelling.

BIM systems have been distinguished on the basis of whether they are design or fabrication oriented, and this is often manifested within the BIM design tools provided. The difference may have a bearing on production systems and interoperability requirements, not to mention the suitability for the user. It follows, therefore, that it is unlikely that a user will to find one particular system or application that is satisfactory for all types of projects, at least at the present time. However, this range of functionality nevertheless remains an aspiration for the future. Some systems interface well with other applications, and some communicate well with certain fabrication software systems, although the fabrication users will, admittedly, be unlikely to require a lot of platforms.

The scale of the task of moving to the use of a BIM system, either as a tool or as a platform, cannot be overemphasised, and the extent of success will depend heavily on early strategic decisions and the associated technologies adopted, not to mention the training necessary to effectively apply it all. In this context it is understandable why the UK Government sought to seize the initiative. Inevitably, over time, with further training and greater integration of the new technologies, the difficulties presented by the new applications are likely to subside; and, simultaneously, the technology itself is likely to encourage an increasing pace of integration, through an accelerating pace in the development of the level of intuitiveness the software's interface.

Interfaces commonly demanded by users of BIM include:

- the ability to scale
- on-demand data generation
- BIM precedents.

An 'open specification', in the context of BIM, describes something that is not controlled by an identifiable single vendor or group of vendors. Such an open specification is the

Industry Foundation Classes (IFC) data model, which describes construction data. It is an often used BIM system. The IFC/ifcXMLSpecification IFC4 is the most recent version at the time of writing (buildingSMART, 2013). IFC was initiated in 1994, and it is the role of buildingSMART (formerly the International Alliance of Interoperability (IAI)), a non-profit industry-led organisation, to publish the IFC classes.

gbXML – The Green Building XML (gbXML, 2013) is an open source that receives funding from the US Department of Energy and seeks to facilitate the transfer of building properties stored in BIM systems to engineering analysis tools. It is said to have the support of most of the major 3D BIM vendors.

4.2.3   BIM as an environment

A single application is ill equipped to be able to process the data generated on a project, particularly as that information becomes progressively more detailed and complex, both within the tools and at platform level. To achieve the status of 'environment' requires the system to possess the ability to generate, exchange, maintain, synchronise and manage a matrix of parametric object data and tools across multiple platforms. In this context, interoperability represents the ease with which the diverse systems within the environment both exchange and use information.

4.2.4   Single model/multiple models

Models can be created for a variety of uses. But a perfectly adequate model may cause difficulties if it is used for a purpose for which it was not intended. Currency, adequacy and tolerances are three issues that need to be addressed when information in a model created for a particular purpose is used in a model created for another purpose.

It seems obvious to state that a model needs to be up to date. A structural analysis model may not need to be absolutely synchronised with the architectural model to determine whether a structure is sound. But the structural fabrication model that can be derived from the structural model must be synchronised with the architectural model, otherwise there will be dimensional conflicts.

Similarly, the detail required in a model depends on its intended use. The end user of the information must understand what information the offered model contains – and also what information it does not contain. Finally, even if the model is current and adequate, the tolerances required may differ between disciplines. The tolerances assumed for structural steel, for example, may differ from the tolerances assumed by a window wall manufacturer. If the tolerances are different, the window wall may not fit when the structural steel is attached. In addition, when performing conflict checking, the models may need to include space around modelled elements to accommodate tolerances or additional material, such as fireproofing.

In its purest form, a BIM project would use a single data model for all purposes. Each participant would access the model, adding content that could be accessed immediately by all others. Exploration, analysis and evaluation would take place within the model, with information being exported as contract drawings, fabrication drawings, bills of materials or other information. However, there are several reasons why this goal is only likely to be partially realised.

For a start, not every participant uses the same software, and not all software is appropriate for all projects or tasks. Designing a software framework that can handle every conceivable project is a daunting task and can result in an overly complex programme. In many instances, modelling software was developed to address issues affecting specific trades, such as piping, ductwork or structural detailing. Not surprisingly, software developed for a specific purpose has advantages when used for that specific purpose, advantages that are not applicable for other purposes. In addition to compatibility, BIM software is often designed to interact with related software, such as structural or energy analysis programmes. Although this approach is very effective if a common engine is used, it can obviously be problematic when merging models that have been built on engines from different software houses. This is one of the major issues that still need to be addressed through the development of BIM.

Thus, because of the above and other reasons, on a single project there are often multiple models that are optimised to a specific task. There are three current approaches to the multiple model problem:

- BIM models are becoming more powerful and capable of handling larger portions of the project. Additional software modules can be added to frameworks to customise the framework for specific uses.
- Standards can be adopted to provide common definitions for the software emulating specific construction elements and systems.
- The third approach seeks to capitalise on the advantages of 'purpose-built' modelling systems and to lessen the difficulties caused by multiple models by using adjacent models constructed on a common framework that are separate but closely linked.

From the participants' viewpoint, the plurality of solutions makes it more difficult to develop a BIM project. Although all the solutions may work, as long as participants are committed to different systems, integration will certainly be challenging.

4.3   A move from the traditional approach

While BIM is not entirely new, the continuing advances in the development of information technology (IT) over the past 40 years have now introduced the growing possibility of a fundamental new development in the methods by which the built environment is both procured and managed. It is fair to suggest that BIM has the capacity potentially to revolutionise those systems in the same way that the main product and manufacturing industries were revolutionised in the last century, first by mass production in the first half of the century, and in the latter quarter of the century by lean production coupled with mass customisation, encouraged within an environment of new and complex delivery methods and quality assurance techniques. But how quickly change actually occurs is uncertain.

The built environment is intended to be created around the needs and desires of its stakeholders, of which there are many. Common standards, tools and processes, combined with a free flow and sharing of information, and a proportionate distribution of risks and rewards, will significantly contribute to the creation of an environment that is conducive to collaboration and the establishment of long-term relationships between its members.

The hurdles that the project participants must overcome, however, are anything but new. In the specific case of drawn information the hurdles consist principally of quality control of the information provided and the method by which the information is circulated among the members of the project team. The data generated within a construction project is commonly instigated by the drawings created in the design process. However, despite the introduction of CAD software, drawings and the information contained within them seem to continue to be a source of friction between project participants with regard to the effectiveness of such documents as a means of communication.

Whenever a drawing is received, all those that receive it must assume that it contains errors. The drawing is therefore checked both for validity and for errors. The drawing also needs to be coordinated with the drawings provided by other disciplines working on the project. The danger for those that do not undertake this task thoroughly and carefully is that they could face financial consequences. This checking procedure takes a lot of time, experience and expertise, one or all three of which may easily be rendered unavailable within the flood of information that is common of a typical construction project.

BIM offers a technologically driven opportunity for built-environment stakeholders to break free of the archaic chains of drawings that bind the process, and to revolutionise the system of design, construction and management of the built environment.

Those BIM systems presently in general use within the UK construction industry commonly follow a traditional approach wherein everything revolves around the lead designer (be that the employer's or the contractor's designer) or, more often than not, the architect, via a system whereby a computerised model of the proposed project is virtually assembled in 3D using precisely formed parametric objects. There are rules that prescribe the order in which parametric objects are created, which include where they may be placed and their relationship with one another, and the objects can be organised to relate to the classes of the components that exist in reality. As time goes on, there will be less emphasis on a 'lead' designer and more emphasis on a 'collective' design approach.

4.3.1 Production rather than project management

In practical terms, the application of BIM possesses the capacity to bear tangible practical fruit in terms of production. In the context of construction management, data on the as-built progress compared with intended progress is undoubtedly crucial and of practical benefit. To achieve this scientifically, rather than intuitively, requires a detailed analysis before commencement, and thereafter continuous analysis throughout the project until completion, and this involves constant measurement and re-measurement of each component under progress of construction, every day. In the traditional craft-based methodologies of the construction industry this form of assessment is far too expensive, too time consuming and too slow to undertake on the majority of projects. However, while the technology is not quite yet able to do so, the automation potential contained within BIM makes such management and control a possibility, particularly as the production plan will be derived from the detailed component schedule compiled by the BIM application.

Traditional construction project management has generally adopted a linear methodology to *progress programming*, with each element of the building executed in sequence to achieve completion (day by day or week by week), while by subtle, yet critical, comparison, a *production plan* indicates instead the level of physical output that must be generated (again day by day or week by week) during the project.

With this in mind, production management overcomes a number of the difficulties associated with project management because:

- intuitive assessments are made redundant
- a direct link is created between records and physical reality
- forecasting and monitoring of progress and production is based on objective scientific analysis rather than being a subjective view
- BIM-based production management provides an objectively derived framework, as opposed to intuition, within which to collate, assess and utilise performance data from one project to the next, providing an opportunity for the development of professional, institutional and corporate expertise.

But BIM is not to be thought of as being a potential panacea for all ills. As with IT generally, BIM is only as good as those who use it and the data that are input to the system. What BIM offers is the possibility for its users to be better at what they do, when they do it. Not only will the use of BIM prevent mistakes being made, but it has the same transformative potential as the IT that transformed the manufacturing and retail industries at the end of the 20th and the beginning of the 21st centuries. If used appropriately, BIM may reduce the number of mistakes made by improving the consistency of the performance of the users, thus affording the opportunity to achieve greater coordination between design and production information among the team. The team will be provided with improved levels of accurate management information, assisting them to identify errors before there are any consequences or to limit the consequences after they occur by identifying the errors in a timely fashion. All of this bestows on the team that is able to harness the opportunities that BIM has to offer the potential for greater profits through enhanced levels of efficiency and cost-effectiveness.

4.4   Incorporating the team

The construction industry consists of an astonishing mosaic of professionals, specialists, subcontractors, main contractors and employers, all of varying sizes and complexity, and all as diverse as the individuals of which they are composed. They all work within a deluge of intensive information flow and under considerable time and budgetary constraints, and are required to produce some of the most complex projects it is possible to conceive.

BIM systems have not yet reached the idealised point of development by which a single designer, using a single system, can design a complete building, and modern projects certainly remain reliant on the involvement of large numbers of professionals and specialists within the overall design, construction and operation stages. However, growing numbers, filtering down from the large to the smaller organisations, are increasingly harnessing BIM or creeping towards the possibility of the ideal use of BIM. They are using sophisticated component-based BIM modelling applications and/or 3D enhanced CAD systems.

There appears to be little doubt that BIM will completely change the way in which designers approach the design process. The level of information contained within a BIM model requires collaboration between different participants at an early stage to establish the initial framework and incorporate the level of detail required. Many more hours will be spent during the design process, and far fewer design hours will be spent in the production stage.

Generally speaking, as a society, we are encouraged at every opportunity to celebrate and pride ourselves on our 'diversity'. In the construction industry, however, diversity is instead negatively labelled 'fragmentation', and conflict, which is a part of everyday life, is mislabelled as 'dispute'. Such 'fragmentation' has formed the basis of various initiatives and studies, both private and high-profile governmental (e.g. Banwell, Latham and

Egan), all of which have sought to address the issues, both perceived and real, arising from such 'fragmentation'.

Regardless of labels, BIM offers the welcome opportunity to achieve greater cohesion, harnessing the benefits of diversity through an improvement in the ease and speed with which those within the construction industry can inter-communicate and share ideas, perceptions and interpretations.

Such ideals, however, depend on interoperability, and in this regard data models are used to manage data and to carry the data. In the context of the built environment, the two main models are IFC (which is used in the planning, design, construction and management of buildings) and CIMsteel Integration Standard Version 2 (CIS/2) (which is used in structural steel design and fabrication) (STEP Tools, 2013). There is also a model for process plant engineering, ISO 15926, 'Integration of life-cycle data for process plants including oil and gas production facilities', which is beyond the scope of this book.

Notwithstanding the diverse, or fragmented (depending up your interpretation), nature of the construction industry, the process of design and construction always has been, and remains, a team activity. The difficulty that this presents to the implementation of BIM is that each of the specialists areas, whether professionals or contractors (in all forms), have tended to develop its own software applications in isolation. As a consequence, there are many ways in which a project may be represented, including in terms of time, cost, geometry and fabrication.

If the team is to collaborate effectively and participate successfully within a BIM environment, it is essential that the system not only dispenses with the need to input data manually but also achieves interoperability between the various applications being used, so that the product of the multiple applications may contribute effectively to the project. The removal of the need to input information manually is a subtle but significant factor in the successful adoption of BIM applications, as there is a direct relationship between the ease with which iterative design is encouraged by technology, and the creation and development of new paths of process automation that ultimately lead to commercial gain.

What does this next stage in the evolution of '3D BIM' require? Certainly it will not take one 'all-knowing' expert, who has won the 'war of the BIM'. In fact, the reality is quite the opposite. More than ever in the BIM process, it looks as if it will take a village of BIM-enabled collaborators who can use the intelligent tools to cope with the pace and quantity of information. In a 3D BIM approach, model reviews, virtual 'huddles' and electronic computer-aided virtual environments (CAVES) will change the environment, duration, nature and results of the process. Shop drawings might be waived in favour of shop models or computer numerically controlled (CNC) fabrication models. Requests For Information might become obsolete, or at least significantly reduced in number, and be resolved much quicker if the model is deployed as a jobsite tool.

Design teams must recognise the benefits of sharing all available electronic information with the entire project team. Structural analysis models, for instance, have value to other team members, so delivery of these models should be part of the design contract.

Along with the responsibility of sharing information, the designer has the obligation to convey the quality of the information that is provided. If the geometry or the load cases in a design model are not completely accurate, this needs to be made known and documented. In addition, the source of the correct information in the design documents needs to be established.

Design teams must also honestly evaluate the submittal process, and work with the rest of the project team to develop the best process for the project. Together the team must find appropriate ways to facilitate communication, without unduly burdening any single member with additional liability. Requiring printed shop drawings or resisting the requests of the team to distribute electronic files, simply because that is the way business has traditionally been conducted, is not helpful to the project or the growth of the industry.

Subcontractors are responsible for fully conveying their interpretation of the design intent to the design team. They also must coordinate their work with that of other subcontractors by sharing the electronic information they have developed, in file formats that can be used and combined with the work of others. They must encourage their software vendors to develop file formats that can be readily exchanged between the various trade subcontractors. Subcontractors also must ensure that all parties understand what they will supply as part of their contract and what will constitute additional work.

BIM is a tool that will help the project team to communicate the needs of the project more quickly and accurately than through current practices. However, the tool cannot perform without the cooperation of the entire team. Each member must contribute its information to the BIM for the betterment of the project, and understand the quality of the information that is included in the BIM.

Current practice is evolving such that contractors and construction managers are taking a lead role in coordination modelling. This makes sense, as they have the most immediate need and can reap the highest return for being able to virtually assemble and view the various components of the project prior to construction in the field. However, each project is unique, and the implementation of BIM should be tailored to the needs of the project. We must remember that BIM is only a tool. BIM will not create, correct nor prevent errors. However, BIM will help find and more fully expose errors earlier in the construction process, particularly when the project team members work responsibly together.

From the users' perspective, the hurdle presented by the difficulty to achieve multi-platform interoperability is that fully automated multiple representations of the same project (e.g. the architectural representation compared with the structural steel work representation) will appear to remain out of reach. Instead, users will continue to rely on manual data transfer between different platforms. The consequence is that changes in one model will need to be studied fastidiously so that those working in other platforms can make the necessary changes to suit the changes made in another model, and vice versa, at great expense to time. Accordingly, full interoperability is critical to the construction team and to the success of BIM.

The stakeholders in the built environment will determine whether and when universal rules will be established, and in what form, all of which is essential to a greater understanding of each other. This is not a matter of translation. The technological language has already been provided to make platform interoperability feasible, but it is the built-environment stakeholders with their expertise (although how demanding each will be will depend on their varying complexity) that possess the knowledge about what must be exchanged between platforms (i.e. as a group they must create and agree on an exchange standard(s), as none of them alone has sufficient weight to establish such rules). Thus the process of creating interoperability includes a process of charting the model information within two applications such that synchronisation may be facilitated between opposing data. Some links may be obvious but others may require the expertise of a relevant construction professional for the appropriate interpretation to be made, hence the above reference to the need for the expertise of stakeholders in undertaking this task.

To achieve interoperability generally, software providers may provide links or exchanges to specific software, and in doing so the same software companies will then be able to provide the necessary support tailored to a certain set of specific needs.

Within the construction industry we have traditionally relied on the creation of drawings to explain and understand the work to be executed. It was always necessary to depict objects with multiple views in sufficient detail to facilitate construction, rendering drawings prone to revision errors and misunderstandings. In addition, drawings are also not easily handled by computers.

Initially, it may be that for those working within smaller companies their earliest experience of BIM will be when they are asked for design information for the BIM model or, if the company possesses a degree of BIM capability, to provide a BIM model in a specific form or manner. At the most basic level, the earliest involvement in BIM may be when receiving a simple request to work in accordance with a predefined COBie data drops spreadsheet (as outlined in Chapter 3).

REFERENCES

buildingSMART (2013) *IFC4 Release Summary*. Available at: http://www. buildingsmart-tech.org/specifications/ifc-releases/ifc4-release/ifc4-release-summary (accessed 7 November 2013).

gbXML (2013) gbXML, version 5.10. Available at: http://www.gbxml.org/index.php (accessed 7 November 2013).

STEP Tools (2013) *CIS/2: CIMsteel Integration Standard*, Version 2. Available at: http://www.steptools.com/support/stdev_docs/express/cis (accessed 7 November 2013).

BIM in Principle and in Practice
ISBN 978-0-7277-5863-7

ICE Publishing: All rights reserved
http://dx.doi.org/10.1680/bimpp.58637.059

Chapter 5
Managing BIM

5.1 BIM protocols and standards

The building information modelling (BIM) protocol and the BIM implementation plan are the key documents that set out the lines of responsibility for the production and coordination of the design throughout the BIM process.

The protocol is the essential foundation upon which successful project appointments, professional indemnity insurance cover and contract documentation are supported. The purpose of the BIM protocol is to maximise production efficiency through adopting a coordinated and consistent approach to working in BIM. It is also used to define the standards, settings and best practices that ensure delivery of high-quality data and uniform drawing output across an entire project. The BIM protocol is essential to ensure that digital BIM files are structured correctly, to enable efficient data sharing while working in a collaborative environment across multidisciplinary teams, both internally and in external BIM environments.

Every BIM project must have its own protocol, and, because of the unique nature of construction projects, each protocol is likely to be unique. The protocol document is the key to achieving a successful BIM project.

A BIM protocol can range from, at its simplest, a model assembly diagram and confirmation that the BIM computer aided design (CAD) standards of a particular BIM participant's office are to be adhered to, to a detailed breakdown of project standards that provide for the development of a shared model for use by other consultants, construction modellers and facilities managers.

BIM protocols set out the ways in which each project has its own unique organisation and structure. They provide the project team with a road map to understanding the aims and objectives of the BIM for a particular project, its rules and how it is to be assembled. BIM protocols provide new team members with the means to comprehend and participate more easily in what is often, by necessity, a complex model structure.

The drafting of the protocol will usually be the responsibility of the BIM manager who, in collaboration with the various BIM CAD managers, will ensure that an acceptable protocol is drawn up.

The protocol is not intended to restructure contractual relationships or stand as a substitute for a complete building project agreement. It is simply an addendum to be appended to the building contract and the various consultants' appointments. Despite this, the protocol addresses: important design, data and process issues, which must be determined at the outset of a project; intellectual property rights; the level of development (level of definition) of the model; model management; the allocation of risk; ownership; permissible uses of the model; schedule of BIM deliverables; etc.

A typical protocol document may (should) contain:

- An introduction to the project.
- An overview of what BIM will be used for on the project.
- The place of the BIM protocol in the priority of the contract documents.
- A reference to any other documentation that should be taken into consideration (e.g. the applicable CAD standards).
- An organogram (Figure 5.1) that shows in a simple way all the parties contributing to the BIM, their obligations, their roles and responsibilities, the delivery time scales and their contact details.
- Details of to what extent will there be a collaborative working practice.
- Details of the BIM manager and who he or she should be appointed by.
- Details of the BIM CAD manager for the participants.
- Definitions of the terminology used in the protocol document.
- Details of the software to be used.
- A matrix showing the different software used and the methodology by which file exchange between each type of software has to be provided.
- Details of the handover procedure from one consultant to another (e.g. where a BIM is moving beyond (existing) RIBA Stage D it is likely that layers modelled by the architect will need to be handed over to other consultants).
- A model assembly diagram (a key, and often forgotten, component of all protocol documents), which allows each team member to understand how the BIM model is arranged. This diagram (or series of diagrams) will have been developed by the BIM information manager, the appropriate BIM CAD manager, the team model manager and the project leader before the protocol is written up.
- The project file structure – if the project is shared, the file structure for the project and the security protocols need to be clearly set out.
- Details of the data security and back-up. The standards for data security and back-up, when record copies of the BIM data are to be retained, and to whom

Figure 5.1 An organogram, which details all the contributing parties, their contact details, obligations, roles and responsibilities, and the delivery time scale. (a) Design, bid and build

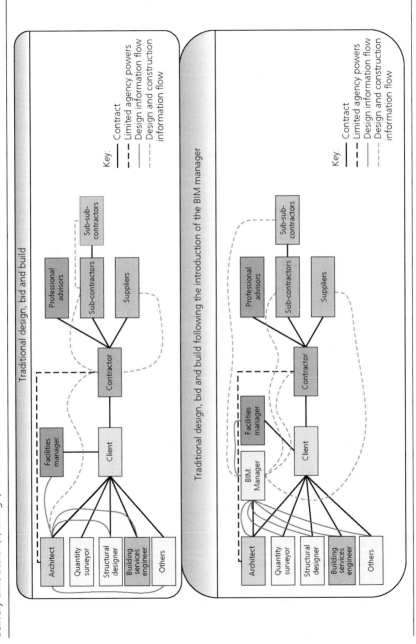

Traditional design, bid and build

Traditional design, bid and build following the introduction of the BIM manager

Key:
Contract
Limited agency powers
Design information flow
Design and construction information flow

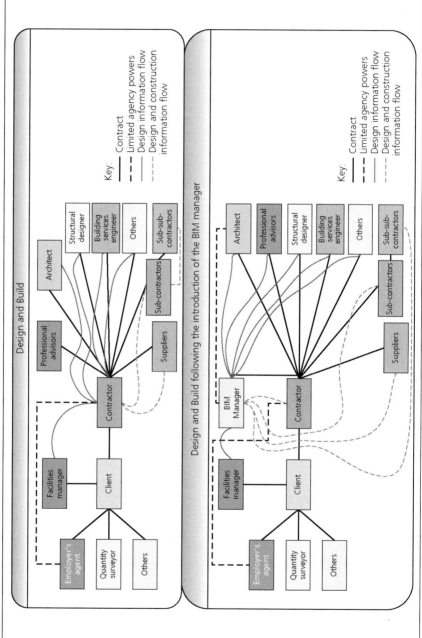

Figure 5.1 (b) Design and build

Figure 5.1 (c) Construction management

Construction management

Key:
— Contract
--- Limited agency powers
— Design information flow
--- Design and construction
    information flow

Construction management following the introduction of the BIM manager

Key:
— Contract
--- Limited agency powers
— Design information flow
--- Design and construction
    information flow

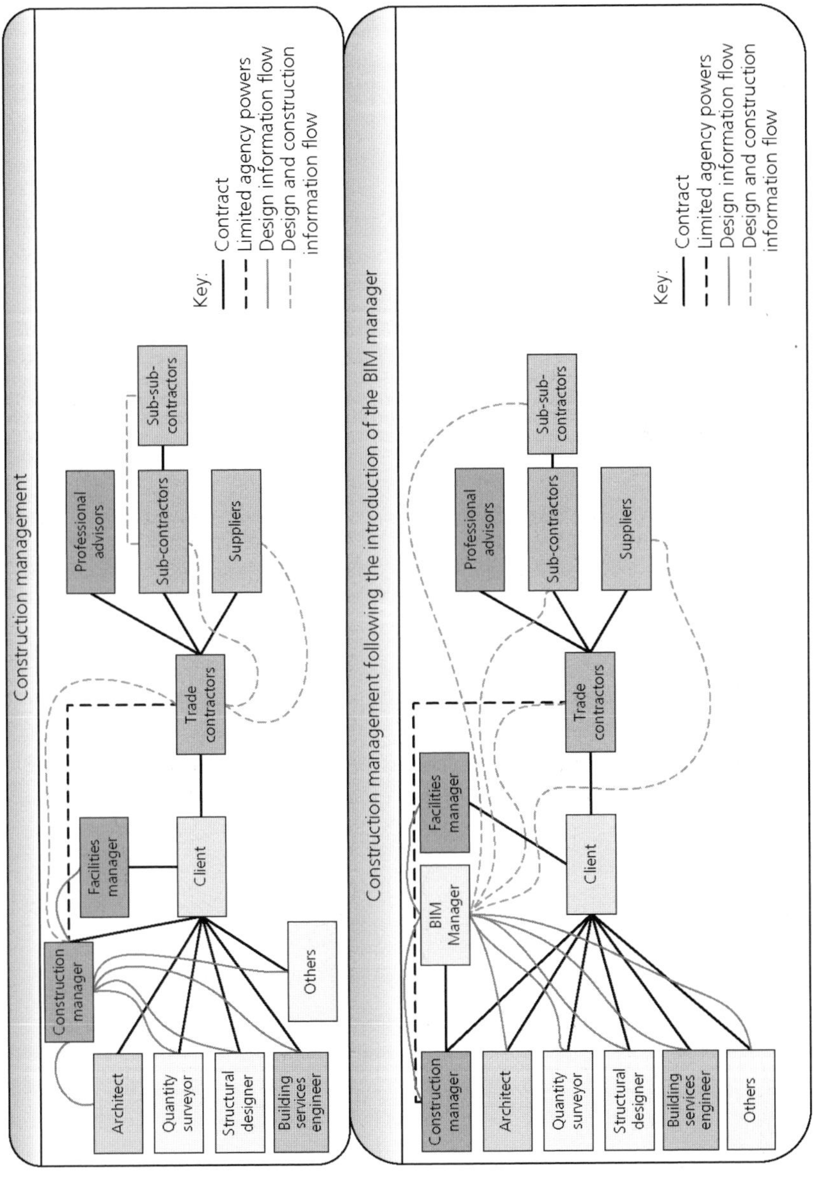

Figure 5.1 (d) Management contracting

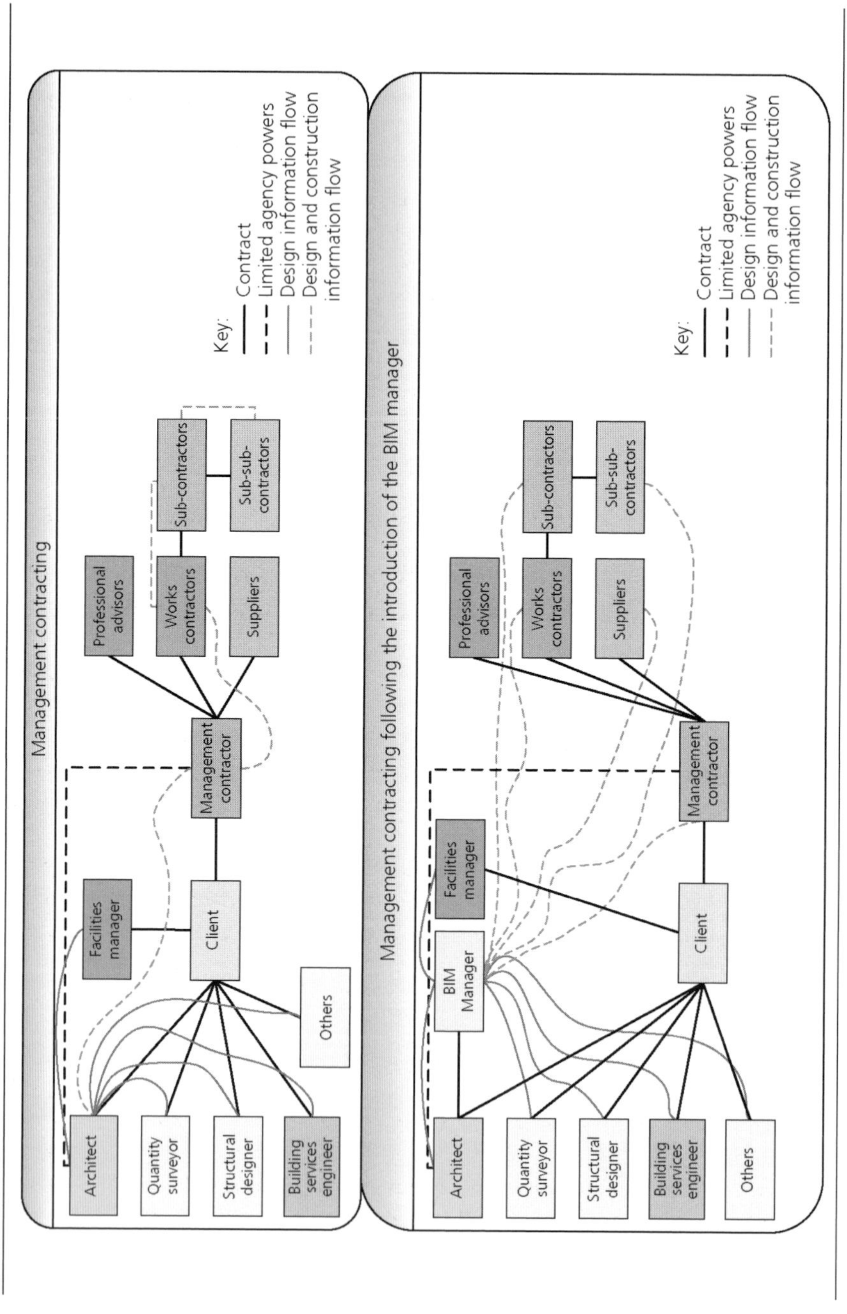

they are to be issued should be set out. The level of security will vary from project to project, with military or government projects likely to have extensive security measures while other projects will require lesser measures. However, it is essential in all collaborative BIM projects that no one contributor to the model is able to alter the work done by another contributor without proper and valid permissions having been obtained.

- The key common data for coordination of setting out. The key data, which all parties need to adhere to, to ensure model compatibility, needs to be stated. These will be, for example, the x, y, z coordinates for the project zero (or confirmation of the use of global coordinates) and floor heights. Note that heights should take into account different consultants setting out preferences, such as 'top of steel' for structural engineers.

- If favourites, new elements and libraries are to be used as laid out as in normal BIM CAD standards then these items need to be stated. If additional data (e.g. cost) are to be linked to elements, this also needs to be stated. It needs to be set out how and where new objects or elements are to be created, and by whom. It must be stated how elements are to be classified, and the library structure for the project needs to be set out.

- If layers are to be used as laid out in normal BIM CAD standards this needs to be stated. If not (e.g. because of a client preference), the layer usage for the particular project in question needs to be set out. The protocol needs to define who has the right to add or remove layers and what the protocol for adding or removing the layers is.

- If pens are to be used as laid out in the normal CAD standards then this needs to be stated. If otherwise, the pen set usage for the project must be stated. The protocol must define who has the right to amend pen sets, and what the protocol for adding or removing them is.

- Change management: the protocol document should set out the method by which change to the model is managed and recorded. BIM enables consultants to work concurrently, rather than following the traditional sequential workflow. Setting out a formal method by which consultants request change will allow change to be tracked and the responsibility for any late change (which may affect the project schedule) to be noted.

- Developing the protocol: protocol documents are unlikely to be complete when first drawn up. There may be a need for additional requirements, or it may be necessary for amendments to be made as a project progresses. Regular reviews of the BIM protocol, both internally and with the other collaborators on the BIM, should be scheduled, and amendments should be made and issued to all parties as necessary.

- The protocol should define who has use of the BIM model(s), who can amend data once they have been incorporated in the BIM model(s), and who can view the BIM model(s) but cannot amend the BIM model(s).

- With regard to copyright, the protocol needs to set out how licences can be granted to other participants for permitted purposes.
- The protocol should also deal with the limitations (if any) on liability associated with the BIM model(s).

The AEC (UK) BIM Protocol Version 2.0 is stated by AEC (UK) to be

'a unification of the guidance provided by the previous documents, bringing workflows together in a single generic document which can be applied to any BIM-enabled project. The set of documents builds on the guidelines and frameworks defined by the UK standards documents, including BS 1192:2007 and the forthcoming PAS 1192-2:2012 alongside proven best practice procedures, providing a clear, concise path to implementation for BIM authoring software, such as Autodesk Revit and Bentley's AECOsim Building Designer.
The AEC (UK) BIM Protocol v2.0 forms the 'hub' of a complete software-based solution. Supplementary documents provide additional detail and enhancements required to implement these protocols using specific BIM authoring software.'

AEC (UK) (2013)

Whether or not the BIM protocol is a contract or contractual document and to what extent the contractor and design professionals can rely on each other's models are important issues to confront and address in the BIM protocol. To have any legal recourse, parties are likely to require that the BIM protocol is indeed a contract document. Most recently, the Joint Contracts Tribunal (JCT) contract amendments, introduced in December 2011 in the Public Sector Supplement *Fair Payment, Transparency and Building Information Modelling* (JCT, 2011) require any BIM protocol to be a contract document. Clause 1.1 in this supplement amends the definition of contract documents, in many of the JCT standard forms, to include 'any agreed building information modelling protocol'.

With regard to the level of reliance on the parties' models (Level 2 BIM), the Consensus Docs 301 BIM Addendum allows parties to choose whether

- each party represents that the dimensions in their model are accurate and take precedence over the dimensions called out in the drawings; or
- each party represents that the dimensions to their model are accurate to the extent that the BIM execution plan specifies dimensions to be accurate, and all other dimensions must be retrieved from the drawings; or
- the parties make no representation with respect to the dimensional accuracy of their models and they are to be used for reference only – all dimensions must therefore be retrieved from the drawings.

In order to avoid complicated and potentially expensive disputes in the future, any BIM protocol should address this dimensional accuracy/level of reliance issue along with the scope of the models created (often referred to as the 'level of definition').

A successful BIM depends on strict adherence to the agreed standards (i.e. standards for software, data storage, data retrieval etc.). Unless there is an unambiguous BIM protocol document to refer to it will be virtually impossible to police adherence to the required standards. Obviously, the policing of standards becomes progressively less of an issue as participants become increasingly more used to working to them. However, when BIM is first used, the required standards must be policed robustly to ensure the successful adoption of the new working methodologies.

Even within a single practice, a model on which many staff are working can quickly become unusable if the agreed standards are not followed. Naturally, when the BIM is to be used by others outside of a single practice it is even more important that the required standards are complied with.

Clearly, when a BIM model is shared with other consultants, the standards agreed need to suit the standards applicable to each of consultants. Because of this need, it is sensible for the standards to be based on a common or universal standards document. One such standard is BS 1192:2007, 'Collaborative production of architectural, engineering and construction information. Code of practice'.

The previous version of BS 1192, dated 1998, provided a guide for the structuring and exchange of CAD data. The new version, with its emphasis on collaborative production, has a clearer focus and has been upgraded to a code of practice. For the first time the new standard offers definitive guidance on how to implement collaborative working, and provides technical details of how to use well-structured names for directories, files and layers. Now it is a code of practice, the design team will need a very good reason not to implement BS 1192:2007.

BS 1192:2007 establishes the methodology for managing the production, distribution and quality of construction information, including that generated by CAD systems, using a disciplined process for collaboration and a specified naming policy. The standard is applicable to all parties involved in the preparation and use of such information throughout the design, construction, operation and deconstruction of projects, and throughout the supply chain. The standard also acts as a guide for developers of applications to enable them to support the implementation of this standard through the provision of configuration files or application add-ons.

The standard has evolved to give professionals a helping hand with how they can manage the information that passes through their hands or computers on a daily basis.

BS 1192:2007 emphasises the importance of effective collaboration between the participants in construction projects in order to enable data to be reused accurately and knowledgeably so that the full benefits can be achieved. The new standard provides guidance to firms in order to effectively share data and enhance the productivity of the whole project team, while also reducing costs.

Collaboration between the participants in construction projects is key to the efficient delivery of facilities. Currently, many projects are increasingly working in new collaborative environments in order to achieve higher standards of quality assurance and the reuse of existing knowledge and experience. A major constituent of these collaborative environments is the ability to communicate, reuse and share data efficiently without loss, contradiction or misinterpretation.

Each year considerable resources are spent on making corrections to non-standard data, training personnel in data-creation techniques, coordinating the efforts of subcontractors, and solving problems related to data reproduction. If the implementation of standards is not adequately addressed, there is the likelihood of significant impediments to both the productivity and the profitability of project teams.

BS 1192:2007 justifies the inclusion of 'Code of practice' in its name due to the much needed inclusion of guidance on the processes that should underpin the creation and management of project data.

Those familiar with the Construction Project Information Committee (CPIC) and Avanti guidance (DTI, 2007) will recognise the BS 1192:2007 guidance as a useful distillation, but may still wish to refer to the supporting CPIC and Avanti documentation. BS 1192:2007 includes recommendations for the implementation of BS EN ISO 13567-2:2002 (BSI, 2002); the international CAD layering standard, but it supersedes that standard and is much wider in scope. The new code of practice is also compatible with the IEC 82045 family of metadata standards, as noted below:

- IEC 82045-1:2001, 'Document management – Part 1: Principles and methods': an international pan-industry document-management standard that is perhaps most useful for its exhaustive classification of the many metadata attributes that may be associated with documents (containers of information), a surprisingly high number of which are applicable to construction.
- IEC 82045-2:2004, 'Document management – Part 2: Metadata elements and information reference model': as it sounds, this contains an XML data model, which would probably bemuse you if you are not an IT professional.
- ISO 82045-5:2005, 'Document management – Part 5: Application of metadata for the construction and facility management sector': this contains modest and

down-to-earth recommendations concerning the metadata attributes of most interest to construction industry professionals.

The take-up of old versions of BS 1192 was surprisingly high. If the new version is equally well received, aside from better processes, there may be greater consistency in file naming within the industry. Naming files according to the standard permits other project team members to determine what is likely to be included in a file, or other named container of information, simply by interpreting its name. When composing a file name, name fields are populated with codes taken from a list of standard or project specific codes and a hyphen (-) is used to separate these field codes. Clearly, if a document management system or project extranet is employed, its facilities for assigning metadata would also be used, but there are many good reasons to adopt a sensible and consistent naming strategy.

Some will argue that the BS 1192:2007 code is meant to support large multi-disciplinary project working (which it is), but much of it is equally applicable to projects of any size, and has evolved and been tested over many years.

The use of IFC or XML file exchanges in workflows began simply with a pair of applications communicating, where a user would be responsible for managing the versions of a project, so that when, typically, the architect issued a revision to and further information on the design, this would be passed to others involved in order to reconcile the information and synchronise their model accordingly. As the complexity of projects and users increases, such coordination becomes less effective, particularly when further ingredients such as time and cost are added to the mix of factors, and it is very likely to cease working if, as with parametric objects, it is objects that are to be managed rather than simply files.

The technology required to overcome the difficulty presented by managing objects rather than files and the need to manage information between multiple applications is a 'building model repository'. Such repositories are server or database systems that facilitate the organisation and coordination of projects at parametric object level, as opposed to file level; this offers the ability to manage data from a variety of applications. With different requirements to those systems in manufacturing, the technology required to support BIM repositories remains a relatively new area. However, BIM repositories will become an increasingly common technology for managing future BIM projects.

The British Standards Institution (BSI) B/555 Committee has been identified as the key producer of guidance, together with a third party, potentially the Construction Project Information Committee (CPIC). It is tasked producing the guidance required to deliver

Level 2 documentation, and details the British Standards (BS), International Standards (ISO) and other standards available and under development.

In the UK, a Publicly Available Specification (PAS) is a sponsored fast-track standard driven by the needs of the client organisations and developed according to guidelines set out by the BSI. In the context of BIM, the draft British Standard PAS 1192-2:2012 is dealt with in Chapter 3 of this book.

5.2　The role of the BIM information manager

The BIM information manager is sometimes referred to as the BIM model manager, the design coordination manager or the VDC (virtual design to construction) manager. The role of the BIM information manager is to:

- provide a focal point for all information modelling issues on the project
- ensure that the constituent parts of the project BIM model have been approved and authorised as 'suitable for purpose' before sharing and before issuing for approval
- ensure that the constituent parts of the project BIM model are compliant with the master information delivery plan (i.e. the delivery plan for the entire project).

Therefore, the BIM information manager's role will include having the responsibility for user access to the project BIM model (or models) and for coordinating the submission of the individual designs and integrating these into the project BIM model (or models).

The BIM information manager should maintain records of who submitted what data and when, and record whether the data were submitted according to the required specification and to the agreed programme. The BIM information manager should also be responsible for data security and for creating a data archive.

Although there may be several subsidiary models in existence, it is during the process of coordinating the data, which is led by the BIM information manager, that the models are usually linked (or referenced) together into one federated project BIM model. Despite this, the role of the BIM information manager is not intended to be that of lead designer, because the manager is responsible for the management of information, information processes, compliance with agreed procedures and coordination of data – not with the coordination of the design itself. If the parties agree that the role of the BIM manager is as outlined here, this needs to be clearly identified and dealt with in the BIM protocol – otherwise there is a potential conflict with regard to who is responsible for design, particularly the design coordination role.

Based upon the above, it is clear that the BIM information manager is a key player in the successful implementation of BIM on a project. Therefore, the BIM information manager needs to hold a senior management position so as to ensure that he or she has the necessary authority and full leadership support.

In addition, the BIM information manager must be a clearly defined and dedicated position, and needs to be placed with someone who has a keen interest in BIM and who will see the BIM process through from the beginning to the end of the project. The BIM information manager needs to be empowered with the authority to make decisions toward achieving the BIM objectives of the project, and needs to provide regular status updates.

The BIM information manager may need to establish roles in the project team with regard to who is doing what and in what format. Assuming that there are multiple models, which will often be the case, the BIM information manager would normally be required to:

- determine who will assemble the composite model
- establish exchange methods for files and formats
- locate a central, shared repository for the drawings
- establish the origin point for model alignment
- implement regularly scheduled reviews and quality checks of the model.

Clearly, regardless of who is maintaining and updating the model during design and construction, someone has to take this responsibility, and it is preferable, by far, for one person to have this separate and dedicated role.

Normally, during the design phase the model will be updated as each consultant organisation issues its next release of design information. However, during construction, the entire project team, including the contractor and the major subcontractors, need to be involved, and be aware that the model is being maintained and updated. Therefore, the incorporation of design changes into the BIM model must be an identified and an identifiable process.

A BIM information manager who is familiar with the various perspectives of and has a well-rounded knowledge of construction would obviously be best placed to yield the greatest return to the project. Such a manager is looking at the overview of the project, and therefore needs to have a big-picture perspective.

A major factor to bear in mind is that BIM is a process that involves team members from several different backgrounds (architects, engineers, contractors etc.), and a successful BIM Information Manager will therefore need to understand how each team member will interact with the 3D BIM project model, and be able to empathise with each of these separate roles.

For a BIM information manager to truly empathise with the various roles, he or she ideally needs to be someone who has a background in one or more of the key consultancy fields (architecture, engineering etc.) or someone who has extensive experience in the construction or civil engineering industry, and also be somebody that understands the technical side of BIM. Therefore, this is quite a demanding role.

However, the BIM information manager should not be overly involved in the technical aspects of BIM, as the challenge faced is to harness technical computerised knowledge to the application of BIM and relate this to the bigger picture.

When it comes to communicating and collaborating with team members, a BIM information manager is really very little different to many other managers; he or she must be a good communicator and this can go a long way to ensuring that everyone is pulling in the same direction. A BIM information manager needs to be diplomatic, yet firm; and, like most other managers, needs to strike a balance between being personable and well liked, while at the same time motivating a team to complete the mission on time and to the required standards.

Because BIM is likely to affect every single aspect of a project, the BIM information manager needs to be a good teacher and an ambassador to train others in the benefits and operation of the BIM process, particularly bearing in mind that the project participants are likely to be at greatly different levels of understanding, and many of them may be afraid of change in any event.

Although there is certainly a steep learning curve when making the leap from 2D to true BIM, the payback is in the potential for acquiring more work, making fewer mistakes and increasing productivity. At its core, BIM is about nothing less than a complete process change. It opens the door to delivering fully integrated projects in which the team shares the knowledge, the risk and the reward more equally. Appointing a BIM information manager is a major step in this process.

5.3   Project BIM coordinator

Sometimes the BIM information manager is also appointed as the project BIM coordinator, as the two roles are very closely linked and in some ways can be considered as being suitable for one person to carry out.

The roles and responsibilities of the project BIM coordinator include:

- the development, implementation and maintenance of the BIM protocol
- ensuring that all stakeholders are in alignment with the BIM protocol

- creating and maintaining a BIM coordination programme that is aligned with the project programme
- identifying any impact on the BIM coordination programme arising from errors in the transmission and use of information during the BIM process
- establishing BIM coordination workshops and reporting progress at the project design team meetings
- keeping a record of BIM models and their status
- keeping a record of transfers of element ownership (e.g. elements of the design work transferring from the architect to the structural engineer, or vice-versa)
- establishing quality control procedures to check that all the models are accurate, and that the level of detail is fit for purpose
- identifying and documenting clashes between different discipline models using clash-detection software
- ensuring that each organisation has published a version of the model for each significant milestone stage as identified in the BIM coordination programme
- recording and monitoring shared data and relationships between models (e.g. grids, floor levels, shared project coordinates)
- identifying and agreeing any co-located or shared technical infrastructure needs, software package interoperability requirements and standards to be used by each team member to deliver the BIM project
- administering the agreed document and model sharing/publication system
- having the exclusive responsibility and power to issue binding instructions on BIM-related issues
- coordinating the handover of the model and data at the agreed milestones in the BIM coordination programme.

5.4   BIM coordination programme

The project BIM coordinator will be responsible for creating and maintaining a BIM coordination programme which is based on achieving the milestone dates in the design team project programme. The coordination programme will define key model release dates, BIM coordination meetings, documentation release dates and any other milestones pertinent to the management of the BIM process.

A BIM coordination programme is essential to ensure that deadlines are met on a BIM project. The collaboration process, especially on fast-track projects, requires careful consideration of the transfer of information in order to ensure that outputs are coordinated. Underestimating the requirements of each participating organisation to reach milestones usually ends up with additional work being undertaken by some organisations in order to ensure that the outputs are achieved, and this leads to frustration. A BIM coordination programme helps the project team understand the effort required and time constraints for the project.

By using and applying a BIM coordination programme, the project BIM coordinator can also flag up with the project manager as early as possible if there are BIM-related delays, and make plans to eliminate or reduce these. This is essential, as BIM is all about coordination.

REFERENCES

AEC (UK) (2013) AEC (UK) BIM Protocols v2.0. Available at: http://aecuk. wordpress.com/2012/09/07/aec-uk-bim-protocols-v2-0-now-available (accessed 7 November 2013).

BSI (British Standards Institution) (2002) BS EN ISO 13567-2:2002 Technical product documentation. Organization and naming of layers for CADConcepts, format and codes used in construction documentation. BSI, London.

DTI (Department of Trade and Industry) (2007) Avanti. Available at: http://www. cpic.org.uk/publications/avanti (accessed 7 November 2013).

JCT (Joint Contracts Tribunal) (2011) *Fair Payment, Transparency and Building Information Modelling*. Public Sector Supplement. Available at: http://www.jctltd. co.uk/docs/JCT%20Public%20Sector%20Supplement%20Dec2011%281%29.pdf (accessed 7 November 2013).

BIM in Principle and in Practice
ISBN 978-0-7277-5863-7

ICE Publishing: All rights reserved
http://dx.doi.org/10.1680/bimpp.58637.075

Institution of Civil Engineers

publishing

Chapter 6
Design liability and ownership

6.1   What is the design?

Traditionally, a design for a building was based on drawings and specifications produced by an architect or another designer. The design may have included input from consultants (e.g. structural engineers, mechanical and electrical engineers) and sometimes from specialist contractors and subcontractors. However, when operating under building information modelling (BIM), the design processes will need to be much more fluid and collaborative.

Certain design elements, such as specific object properties, will be created by suppliers, specialists or software manufacturers, not by what may more traditionally be considered as licensed design professionals. In addition, the design itself may be self-modifying, and to that extent it may be partially self-designed.

The design deliverables are more likely to be a computer model or simulation than paper drawings, and the resulting model may be distributed between computer systems operated by different participants. Participants may also themselves add to a design model, and the design will therefore be fluid and flexible. Furthermore, the complete design (or partially complete design) may exist on a cloud server rather than as a set of hard-copy drawings.

The above concepts bring with them certain challenges. For example:

- At tender stage, the design needs to be clearly expressed and defined, as contractors need to know that they are all bidding against the same basic design information.
- The revised design elements from earlier models need to be easily identified so that variations to the original scope of works can be demonstrated.
- Inspectors and other building officials must be able to compare the physical construction with an objective and stable design standard, not with a model that constitutes a 'moving target'.
- Designers need assurance that their services are complete, and, if problems occur later, that their designs can be compared against the constructed condition.

- Owners need to determine whether they have received a project that complies with the design.

Clearly, the fluidity in design that is allowed for by BIM technology is likely to compete with the precision required for contract enforcement. Therefore, the contract definition of design needs to address two key issues:

- The contract documents between the various parties should define the design deliverables in terms of content, time and type of electronic media to be used.
- The contract documents between the various parties should determine whether incorporated submittals, such as parametric objects provided by suppliers, specialists etc., are part of the designer's deliverables, and which party is responsible for the incorporation and coordination of the submittals.

Once a design definition has been adopted in a contract, it will be important for the parties, and particularly for the designer, that the definition is adhered to during the project development.

For the ease of future reference, the design should be preserved as 'snapshots' at major design milestones. In some cases this may be accomplished by printing and saving the milestone documents. However, in a multi-dimensional electronic design maintained in a diffuse internet relationship, the total design package may never be encompassed within printed documents. In such cases it may be possible to temporarily freeze the digital design at a particular time, and save it complete with linked documents and locations on semi-permanent media such as CD-ROMs.

In the meantime, and in parallel, the model containing the design will continue to be developed by the various approved participants.

6.2 Who are the designer and delegated designers?

As outlined above, when BIM is applied, not only is the concept of 'the design' less clear, the identity of 'the designer' is equally vague. In all probability there will still be a designer who is responsible for the overall design concept, and who will maintain responsibility for systems design, the overall layout of design elements, the flow through the structure, and the 'artistic' elements of the building. However, there will also probably be many other 'designers' (including consultants, suppliers, specialist contractors and subcontractors) who will add to the design. All these various designers will now be obliged to work together in a collaborative way.

In such a collaborative setting, where the design input can easily be 'hidden' in a fluid design model process, several significant questions are relevant:

- In the case of design failures, how will the various designers' contributions be extricated from the model to determine responsibility for the design failure?
- Will parties accessing the shared model be legally able to rely on the contributions of other designers, or will the legal doctrine of privity of contract (i.e. where only the parties to a contract owe duties to one another and realise any benefits under the contract) become an issue?
- What are the responsibilities of the secondary designers?
- To what extent can the design professional rely on the products of the secondary designers?
- If the software can communicate between objects and cause them to adjust their properties, does the software itself become a 'designer' as well?
- Do the standards committees that develop interoperability protocols and object specifications also become project 'designers'?
- If these other parties and entities do have design responsibility, do they have insurance for the design risks?

In the immediate future, owners and building users and developers will look to the architect and engineers of record as being the project's designers. But, in a practical sense, these parties cannot check and be responsible for the work of the many other individual designers distributed throughout a collaborative (and possibly convoluted) design process. Most disputes regarding design deficiencies have little to do with the main design elements, but rather arise from deficiencies in details, inadequate coordination, deviations in submittals, excessive changes, and failure to meet budgetary or functional programme requirements.

Just as tomorrow's designs will be distributed between many parties, it is clear that so should design responsibility. Therefore, in developing contract documents, careful thought should be given to integrating appropriate limitations of liability and waivers.

The professional registration statutes generally require that a licensed professional be in 'responsible charge' of all work performed by a design firm. This work must either be performed by or supervised by the responsible professional. The contract documents are sealed by the responsible professional to signify compliance with this requirement and acceptance of this responsibility. If design responsibility is distributed, however, is this even possible? How can a professional supervise design contributions by firms that are not under his or her control? How can a design professional supervise changes to structural detailing that are performed by the software itself?

In the short run, building officials are likely to accept sealed drawings without considering what portion of the content has been created under the responsible charge of the signing professional. However, in the long run, the professional registration statutes must be modified to reflect the actual practices, and realities, of digital BIM design.

Design delegation creates potential issues with licensing and with responsibilities. BIM designs, especially when based on the design of specific objects that have encompassed within them their own design parameters, can contain embedded information provided by manufacturers and subcontractors. In addition, some BIM software can react to changes in the model. Structural design software, for example, can change details in response to changes in the design. In neither of these cases will the architect or the engineer have created the information, and probably will not even have been able to check the information before it is incorporated in the BIM model.

Naturally, architecture and engineering practice will continue to evolve and use increasingly powerful design tools, but the legal and regulatory structures have not yet fully adjusted to this change in practice. These liability issues highlight concerns that arise from the distribution and delegation of design. Although design delegation issues can exist with or without collaborative use of BIM, they are clearly much more significant when more parties are involved and are involved more closely with each other.

6.3   The designer's liability

As noted above, in a collaborative BIM model many parties contribute to the design. Crucial details embedded in the design may be provided not by design professionals but by specialist subcontractors or consultants. In addition, BIM software is designed to react to changes in the model, by modifying elements of the design affected by a change. These circumstances increase the potential liability exposure of design professionals who use BIM collaboratively, and they risk assuming overall model responsibly.

This can lead to concerns about whether or not the use of BIM might alter the traditional allocation of responsibilities between the client, contractors, designers and suppliers. If each party is responsible for its own model, to what extent is the BIM information manager liable when clashes are not detected or the design is not coordinated?

The typical approach, at least in common law, appears to be that set out by the draft PAS 1192-2:2012 (BIM Task Force, 2012), which suggests that the lead designer shall be responsible for the coordinated delivery of all design information. In other words, nothing has changed.

The role of the BIM information manager is, therefore, not meant to be equivalent to lead designer. The BIM information manager is responsible for the management of information, information processes and compliance with agreed procedures, not the coordination of design. However, this does need to be spelt out, perhaps in the BIM protocol; otherwise there is potential for conflict with regard to the design and design coordination roles.

In the UK, where the government is proposing the implementation of Level 2 BIM, it appears to be the case that BIM should not alter the traditional responsibilities to any great degree. This is because Level 2 BIM is a series of federated models prepared by different design teams (the number of models and their purpose is determined by the employer) and put together in the context of a common framework for the purpose of being used for a single project, with licences granted to other project teams' members to use the information contained in the federated models. Therefore, if each model is thought of as a drawing or design in the more traditional sense, then, provided that the individual contracts clearly define the role and responsibilities of the designers in the usual way, there should not be any significant change.

Although the usual responsibilities will remain, and it will be important that the design brief is understood and that the design is reviewed as an ongoing obligation, the new BIM technology and the new way of producing design should not, in themselves, change the fundamental legal principles. Consequently, there will still be the need for a lead designer to coordinate the design.

Therefore, BIM will not be a cure for all evils, but if BIM reduces drawing errors and miscommunications between the parties, the frequency and severity of loss will be lessened, even if the number of potential reliant and liable parties is significantly expanded.

6.4 Information ownership and preservation

A dynamic model creates challenging issues around ownership and preservation. The model is immensely valuable but can be fragile. Computer software is susceptible to power interruptions, viruses and physical damage.

An essential element woven throughout the vision of transformation to an optimised model is the ability of all parties to communicate freely. Current practices of silence for fear of liability must obviously be eliminated, and a new process where decisions are made at the highest and most appropriate level of competency must be established to leverage team knowledge. This issue most certainly is the greatest obstacle to transformation and the realisation of the optimised project. Owners must demand this openness and transparency from the team entity of which they are a part.

Although these dangers can be reduced by appropriate back-up strategies, there are risks involved with hosting data, and BIM models, like all digital data, are susceptible to data loss, and even small data losses can require significant effort of recovery or replacement.

Therefore, if a party is hosting the information, it must take adequate steps to protect, and insure, against data loss, or it will face possible liability for the ensuing losses (which could be very significant in size). A design firm can purchase 'valuable papers' coverage that provides catastrophic loss protection, but this will not necessarily cover losses to other collaborative users.

If a failure occurs, what insurance, if any, will respond to the economic losses? Coverage under the designer's professional liability policy is problematic, and the designer's commercial general liability policy will not cover purely economic losses. The difficulty in characterising and insuring against this type of loss underscores the need for comprehensive risk allocation and waivers among all BIM model users.

Data preservation can also be challenging. We have recently seen extraordinary judgments and sanctions levied against corporations that did not appropriately preserve relevant electronic evidence. The duty to preserve evidence arises when litigation can be reasonably anticipated. On a construction project, however, claims are a normal aspect of project close out, with only some claims proceeding to litigation. Unfortunately, when they do arise, claims that are eventually resolved by the parties look strikingly similar to claims that result in litigation. After litigation commences, the likelihood of litigation will look 'reasonably anticipatable' in hindsight.

Even assuming that the design professional could recognise when information needs to be preserved, it is unclear how that should be accomplished. An advantage of a dynamic model is that it can and does evolve. This inherently involves replacing information with newer information, and overwriting or discarding the obsolete data. Although systems can track revisions, they may not be able to accurately roll back every change made to the system.

Moreover, the BIM model differs from traditional paper documents (or even electronic word processing files) in that there is no single paper representation of the model, and critical information is contained in the relationships between information. It is the BIM model, and not its manifestations, that need to be preserved.

In most projects, ownership of the BIM model is likely to be retained by the owner of the building. However, ownership of the data contained within the BIM model itself is a separate issue. Such data are likely to be wide ranging, and contributions will come from a variety of different participants. For example, there is likely to be design data, cost data, design processes, tables, databases and graphical information. Different laws may well govern each of these rights.

Enhanced attention to intellectual property provisions is especially important, because the BIM model will not only show the results of patented processes and designs but will actually 'know' the building codes, algorithms and applicable engineering principles. It is this information which the system applies to enable the model of the building to be manipulated and updated. Therefore, design participants will be required to impart much more intellectual property on a BIM project than they are traditionally used to. The terms of any assignment or licensing of intellectual property rights will, therefore, be of great concern to all design participants. Increased integration and collaboration also means that there will be complex layers of intellectual property, provided by different design participants, which will be difficult to identify and reverse engineer.

Copyright and intellectual property provisions are an implicit form of ring-fencing information in the agreement. A single island is a single interest that is unlikely to be in the project's best interest or in the collective interest of the entire team.

6.5   Data translation, interoperability, storage and retrieval

As noted previously, there will rarely be a single BIM model on a complex project. The architect may have its design model, the structural engineer its analysis model, the contractor its construction model, and the fabricator its shop drawing or fabrication model. In theory, these models will communicate seamlessly. But under current technology this is an aspiration, not a reality.

Confidentiality provisions should be reviewed for consistency with similar requirements in exchange agreements. Obviously, as the use of BIM becomes increasingly commonplace, standards will be developed defining who is responsible for inputting what information into a model, but we are not at that stage yet.

BIM and other collaborative technology will compel owners, contractors and designers to interact in a different way than they have traditionally done so. As those relationships change, so too will the contractual language that defines them. However, the current uncertainties that accompany the changes brought about by BIM need not inhibit contractors from experimenting with and ultimately embracing the future of construction. The traditional tri-party approach to design and construction will, over time, be replaced by 'integrated practice' and collaboration.

In current practice, there are differences in capability between different types of BIM software, and this is a major hurdle that needs to be addressed. Clearly, information must be translated or must fit into the standards for Industry Foundation Classes (IFC). Translators may not transfer all information from one model to another. In addition, some translators cannot 'round trip', that is move data from one platform to another and then return it to the original platform after it has been modified or augmented. IFC do not exist for all data types. and there can be data loss if the host application supports functionality not modelled in the IFC. The net result is that differences can be created during the translation process itself that can cause model inconsistencies and errors.

The parties should review contract provisions in the design and construction agreements that address the line and flow of communications between the project parties. This allows them to assess whether such provisions need modification in appropriate, limited circumstances to permit direct communications between parties not in contractual privity, such as the principal design professional and a specialist contractors or subcontractors performing a portion of the design. In such circumstances, contemporaneous notification of such communications or exchanges should be given to the parties through a particular line of communication.

Methods must be established for maintaining version control of electronic documents, including a depository of record copies of transmitted and received electronic documents. Contractual reporting requirements for known or observed errors or omissions in contract documents should be reviewed, to ascertain whether they are adequate and consistent given the potentially increased pace of electronic document exchanges.

At the end of a project, archiving raises technical and practical issues. Although it is possible to save the model onto electronic media, this does not guarantee that the saved model will be usable. Properly prepared paper has an archival life of 100 years and, if carefully preserved, longer. We have limited experience with the long-term reliability of digital systems. We are aware that most magnetic media have a limited lifespan. CDs and DVDs can last considerably longer, but that may be irrelevant.

If data are archived on currently popular media, using currently popular software, it may well be the case that it is difficult or impossible to restore or view the data when needed at a much later date. Therefore, an issue that needs to be considered is how long do we need to maintain models and how should this archiving be accomplished?

6.6   Intellectual property

Given that the intelligent model is an inherently collaborative work, to what extent can anyone claim ownership of the intellectual property? In select instances, the designer's intellectual property rights have been used to preserve the integrity of the design itself. More commonly, the intellectual property rights are used to enforce payment obligations or to prevent reuse of the design without compensation. Because the client will ordinarily have access to the model as it is being developed, care must be taken to ensure that the intellectual property rights are not lost because of the open and collaborative nature of model development.

The model may also contain confidential or trade secret information. For example, a model for a manufacturing plant may disclose what a company is planning to build and the processes it will use. If information is broadly circulated in a collaborative team, how will this information be protected legally and practically?

Many of the intellectual property issues are similar to those that existed before BIM. However, they are amplified by the amount of information contained in the BIM, the access to this information by others and its ease of transfer.

At the most fundamental level, the question is one of who owns the information in the BIM model. If the BIM model is a collaborative work, then ownership may not be vested in a single party. If ownership issues are significant, they should be determined by contract. If information is confidential, then care must be taken to limit the distribution of information and appropriate confidentiality agreements should be in place.

Confidentiality issues can arise subtly when the embedded information itself is confidential, although the overall design is not inherently confidential. The upshot is that who owns the model, who owns information in the model, and who has access to the model should be considered when BIM procedures are developed and when intellectual property rights are considered.

REFERENCES

BIM Task Force (2012) PAS 1192-2 Specification for information management for the capital/delivery phase of construction projects using Building Information Modelling. Available at: http://www.bimtaskgroup.org/pas-1192-22012 (accessed 7 November 2012).

BIM in Principle and in Practice
ISBN 978-0-7277-5863-7

ICE Publishing: All rights reserved
http://dx.doi.org/10.1680/bimpp.58637.085

Chapter 7
Contracts

7.1   Collaboration

Building information modelling (BIM) technology can radically change the form of the work product of several members of the project team, and it can allow projects to be built faster, with fewer surprises and with lower costs.

BIM technology can, of course, be used solely to produce better quality design documents without any intent to share information or to use the more extensive functionality that BIM allows. When used in that limited way, BIM is simply an advanced form of CAD, and adds very little to the existing process.

However, in order to optimise efficiencies from a process such as BIM, a collaborative team structure needs to be in place. That team structure needs to be one in which team members are either contractually obliged to, or in some other way have agreed to, work in a unified manner, and the structure needs to be one where the team members will provide each other with data that will allow the other team members to perform their work faster, better and/or cheaper. At the same time, each team member needs to be able to insert, extract, update or modify information in the BIM model, to support and reflect its own role in the process and to ensure that each project member remains responsible for its own element of the design and/or the data input (or not, if that is what has been agreed). Used in this way, BIM serves as a catalyst to change for the relationships between the parties, and eventually for the basis of their agreements.

Collaboration through BIM is a profound change that creates great opportunities, as noted above, but also creates new legal and liability issues. All significant construction projects in the UK incorporate a written contract at their core. At their most basic level, contracts legally record that which has been agreed upon by the parties to the contract. But not all agreements are legally binding. To constitute the 'classic' contract under English law, there are essential elements that must exist and elements that must not exist. Beyond that basic description, construction contracts are as complex as the matters within, and also beyond the original anticipation and the original contemplation of the parties to the contract.

85

It is what construction contracts oblige stakeholders to do, refrain from doing, either expressly or implicitly, or those matters upon which the contracts remain silent that have formed the basis of construction contracts being seen as adversarial.

Contracts determine the parties' risks and rewards, but also the rights and obligations of the parties, and also sometimes the wider stakeholders. Thus the construction contract creates a legal framework upon which the stakeholders' duties and obligations are defined.

There are various standard form contracts available (e.g. the Joint Contracts Tribunal (JCT); the New Engineering Contract (NEC)) and, while some contracts are written bespoke for a specific project, the majority are simply one of the standard form contracts with bespoke provisions and/or alterations to the existing wording to suit the particular requirements of a project and its stakeholders.

For many years, a major concern has been that the adversarial nature of the construction industry has been cultivated, in part, by contracts that are viewed as being confrontational in nature and which can quickly escalate disagreements to disputes. Many initiatives have been launched to consider and deal with this perceived problem, going back to the early 1960s with the Banwell Report. In more recent times, and particularly over the past 20 years, the drive for reform has intensified.

The Latham report, *Constructing the Team*, was published in July 1994. It arose from a commission given to Sir Michael Latham to review the procurement and contractual relationships within the UK construction and engineering project industries. The report raised questions about how construction contracts should be drafted, and important questions with regard to how the industry participants contract with one another. While proposing basic principles of contract drafting, Sir Michael Latham criticised the existing construction industry standard forms (at that time) and the means by which they were produced. The guiding principles arising from the report were for simply worded and fair contracts that promoted team working with clear management processes. Moreover, Sir Michael observed that the existing standard forms of contract did very little to solve the adversarial aspects of the construction process, noting at the time that those forms that were produced by the JCT and Institution of Civil Engineers (ICE) were often heavily amended or not used at all, and that if they are to be used they should be amended to meet the principles that he advocated.

In the wake of the Latham Report, calls for innovative practice, and a reduced dependency on both competitive tendering and adversarial contracting, grew from the public sector, and, following a comprehensive review in 1995, the Levene report made umber of recommendations to improve the procurement and management of construction projects: this included the reduction of conflict through enhanced levels of communication.

It has often been argued in the context of the proposed contract legislation following the Latham report that what was really required was a culture change in the construction industry, because the teamwork changes advocated by Sir Michael Latham simply could not be legislated upon because they would be the product of trust, relationships and mutual understanding, and such matters really were beyond the control of contracts.

By 1998, all the UK construction standard forms of contract had been revised by their respective authoring bodies to reflect the principles espoused within the Latham Report and the legislative changes that followed it (i.e. the Housing Grants, Construction and Regeneration Act 1996).

The report, *Rethinking Construction* (Egan, 1998), produced by a task forced chaired by Sir John Egan, considered the construction industry from the client's perspective with the objective of identifying opportunities for improvements in efficiency and quality. It suggested that contracts may be replaced with performance measurements.

In March 1999, the UK government's *Achieving Excellence in Construction* (Office of Government Commerce, 2007) initiative was launched, and this set out a route map to four improvement objectives: management, measurement, standardisation and integration.

The Strategic Forum for Construction was established under the Chairmanship of Sir John Egan in July 2001, and in September 2002 published *Accelerating Change* (Strategic Forum for Construction, 2002), which emphasised the role that information technology could play in achieving greater integration of construction teams. This formed the basis of subsequent initiatives, the most notable of which, at present, is the drive towards the implementation of BIM.

In May 2013 of the Construction Industry Council (CIC) published the first edition of its *Building Information Model (BIM) Protocol CIC/BIM Pro*.

In July 2012 the Procurement/Lean Client Task Group presented its Final Construction Strategy Report to the UK Government. The report provides recommendations, the first five of which are as follows.

1 Three or more trials of each of the three models should be made available from the public sector.
2 Trials should apply collaborative forms of contract. Cost-led procurement trials should use NEC 3 option C, Integrated Project Insurance should use PPC 2000, and Two Stage Open Book should use JCT Constructing Excellence.

3   In each case contracts should have absolute minimum of amendments, with no changes to risk allocation or payment terms except where they are improved.
4   Effort should be taken to avoid the use of liquidated damages, retentions, parent company guarantees and performance bonds on the trail projects.
5   Client and supplier teams involved in trial projects should be provided with professional development, experiential learning and hands-on training to ensure that they adopt the intelligent client attributes and operate in a collaborative culture as identified in Appendices E, F and G.

(Procurement/Lean Client Task Group, 2012, page 5)

The report also identified objectives, the first of which is:

While much of the reform that is identified in the Government Construction Strategy relates to the improvement and update of existing models and behaviours, there is recognition that to achieve optimum efficiencies the public sector should consider new approaches to construction procurement.

The Government Construction Strategy Action Plan calls for:

- Investigation of alternative forms of procurement and contractual arrangement that offer better value and affordability
- Demonstration of the effectiveness of these alternatives through trial projects.

(Procurement/Lean Client Task Group, 2012, page 12)

The *Government Construction Strategy* made two proposals for possible new models, which should, in its opinion, be trialled: cost-led procurement and integrated project insurance. The group also suggested a two-stage open-book proposal, which had not been identified in the strategy. With regard to the integrated project insurance proposal in particular, the group said it expected overall cost savings of 25–40%, which included a 15–20% saving by the 'removal of adversarial culture'.

Therefore, it is clear that over the past 20 years, there has been growing momentum in the belief that new ways must be established in which the construction industry works together. However, until attitudes and culture are addressed at a much more detailed level, so that a balance is struck between aspirations and attitudes, innovative contractual practice will not become reality.

It is clear that the various construction industry initiatives and commentaries by others reflect the position that stakeholders to the construction industry attribute a considerable amount of the difficulties they face to the nature of the contract. It is often considered that the standard forms of contract (particularly those with ad hoc alterations and amendments) pit stakeholders against one another, rather than effectively managing conflict and promoting the pursuit of amicable solutions when situations arise.

The construction industry is important to the wider UK economy. Simply viewed in financial terms alone it remains obvious that it is in the nation's interests as a whole to maintain a healthy construction industry. The importance of the construction contract simply cannot be overstated.

With all of the above in mind, and while the present focus of the current range of initiatives appears to rely heavily on technological advances to achieve better information management, it could be that, because of the 'enforced' collaborative nature required by BIM, the construction industry has its best ever chance of ridding itself of contracts that breed confrontation, conflict and dispute.

7.2   New contracts

Traditionally, most contracts are bipartite agreements (i.e. an agreement between two parties only) with clearly defined boundaries in terms of scope and liability. These contracts tend to insulate and isolate rather than collaborate. One contract tends to define one pair of relationships or one set of requirements in a project, each having little regard for the other.

Our legal system is essentially individualistic, focusing on individual rights and responsibilities. We expend great effort to determine where the responsibility of one party ends and the responsibility of another begins. Many of the most fiercely fought battles in construction law focus on the dividing line between entities. Privity of contract, the economic loss doctrine, means and methods, and third-party reliance are all issues where drawing lines between parties is essential to determining responsibility and liability.

In the contracts in use today there are very few issues at the Level 1 stage of BIM maturity, as the BIM tools are commonly internal to members and do not necessitate alterations to the contract documents. However, the need to consider appropriate amendments to the presently published (at the time of writing) standard forms of contracts begins to arise once Level 2 of BIM maturity is reached.

Level 2 BIM provides data and information in a 3D environment, with each member of the design (and possibly construction) team creating and maintaining their own models. These models and databases then 'fit' or work together with the use of proprietary technology. This consolidated model, composed of the individual models, prepared by each discipline, is often referred to as a 'federated' model in a common data environment.

The consensus of opinion appears to be that the use of BIM at Level 2 does not require wholesale changes to the traditional forms of contract or the allocation of responsibilities between the parties. The view of the NEC, for example, is that there is no need to do anything more than insert a BIM protocol into the works information document, and this approach is also taken by the leading UK standard contract body, the JCT, which, in its Public Sector Supplement, *Fair Payment, Transparency and Building Information Modelling* (JCT, 2011), suggests incorporating a BIM protocol as a contract document.

In support of this position, in March 2011 the Government Construction Client Group (GCCG) concluded in its strategy paper that (BIS, 2011):

Little change is required in the fundamental building blocks of copyright law, contracts or insurance to facilitate at working at Level 2 of BIM maturity. Some essential investment is required in simple, standard protocols and service schedules to define BIM – specific roles, ways of working and desired outputs.

However, in contrast to the above, Level 3 BIM utilises a single project model, accessible by all team members. Therefore, as BIM moves towards Level 3 in the future, changes to building contracts will almost certainly be necessary, as the traditional legal position and relationship between the parties are likely to change.

The Level 3 of BIM maturity will raise significant contractual, legal and insurance issues, including:

- the priority of contract documents (e.g. between the contract conditions and the BIM protocol)
- multiple model relationships and conflict prioritisation, including data derived from them
- design liability and thereby professional indemnity insurance
- matters concerning intellectual property rights.

Many of the above issues are associated with collaboration (which will be required if BIM is to operate efficiently and effectively) and relate to duties and obligations that transcend boundaries between normal bipartite contracts.

As the use of BIM develops further to the level of fully integrated BIM, it may be that bipartite contracts, even with BIM addendums or protocols, may become unsuitable, and collaborative multiparty contracts could potentially become more appropriate. The BIM process has the potential to substantially alter the relationships between parties and blend their roles and responsibility. Risks will need to be allocated rationally, according to the benefits a party will be receiving from BIM, the ability of the party to control the risks, and the ability to absorb risks through insurance or some other means.

There will be a need for standard contract documents suitable for BIM, and without these the development of BIM will almost certainly be hindered. Unfortunately, the current standard contract documents are only just beginning to address the issue of fully integrated BIM. BIM's implications are just being realised, and so far few solutions have been developed.

Standard contract documents perform four key functions:

- they validate a business model by providing a recommended framework for practice
- they establish a consensus allocation of risks and an integrated relationship between the risks assumed, compensation, dispute resolution and insurance
- they reduce the effort involved in documenting the roles and responsibilities on a project
- the drafting of bespoke documents increases the transaction costs and thus reduces the profitability of every transaction.

Unfortunately, the emergence of BIM as a vehicle for dramatic change in design and construction occurs in a legal environment that has not fully come to grips with all the risk management implications of the underlying technology of electronic representation, or transmission of documents of any type.

It may be said that liability only requires that there be intent to influence and reach a group or class of persons, and therefore contractors and subcontractors relying on the design within a model may be able to bring an action against the designer for damages caused by negligent errors. However, in a collaborative project, the designer is aware that other parties are relying on the model's accuracy, and it is a very short step from foreseeability to knowing that the model is intended to provide information for the contractors' and subcontractors' benefit.

Against this background, it is clear (particularly as the process develops further) that BIM is essentially collaborative, and it will be most effective when the key participants are jointly involved in developing and augmenting a central BIM model. Therefore, although roles and duties will remain, the transitions and boundaries between the various team members will be less abrupt and less easily defined.

As the leaders of construction coordination, contractors and construction managers have a responsibility to encourage and facilitate the sharing and distribution of BIM technology on a project. However, they must also understand and convey the nature of the information that is being shared. Appropriate contract language that will foster the open sharing of BIM information must be developed. The contract language cannot alter the relationships of the project team members or change their responsibilities beyond their ability or what they are permitted to perform. Therefore, there is a tension between the need to tightly define responsibilities and limit reliance on others, and the need to promote collaboration and encourage reliance on information embedded in a BIM model.

But, even as lawyers spot the legal issues, how best to resolve those issues remains an open question. Some fear that an excess of concern over all the potential questions of liability, risk allocation, shifting and sharing associated with BIM might inhibit many from experimenting with it, and in the process deny owners, designers and contractors the opportunity to sort through the issues as they experiment in the laboratory of the real world. Because of this, the terms of engagement for all members of the project team will need to be considered and drafted to reflect the collaborative nature of BIM and to ensure that their responsibilities, duties and services are aligned.

Communications and methods of working will need to be outlined to enable those who are not in direct contractual relationships to work together with others in order to deliver the project in the integrated and collaborative spirit required by BIM. Provisions on the input of data, limitation of liability and defining the role of the BIM 'model manager' will also be crucial.

The alliance approach to contracting has been successfully used for oil exploration, for the delivery of infrastructure and in other fields also, and therefore there is no logical reason why it cannot be used successfully in the construction industry.

Project alliancing may be defined as a commercial/legal framework between a building owner and employer and one or more parties for the delivery of a project. It is characterised by:

- collective sharing of project risks
- no fault, no blame and no dispute between the alliance participants (except in very limited cases of default)
- foresight applied collaboratively, to mitigate problems and shrink risk
- a clear division of function and responsibility, which helps accountability and motivates people to play their part
- payment of parties providing services under a three-limb payment plan comprising:

- the reimbursement of the party's project costs on a 100% open-book basis and/or on a more front-loaded basis
- a fee to cover corporate overheads and normal profit
- a gain/pain share process where the rewards of outstanding performance and the pain of poor performance, and the financial gains and losses are shared equitably among all alliance participants
■ unanimous principle-based decision-making on all key project issues
■ an integrated project team selected on the basis of the best person for each position.

The NEC3, ACA PPC 2000, Strategic Forum's Integrated Project Team Agreement and JCT Constructing Excellence forms of contract are all steps towards partnering or alliancing. The Strategic Forum's Integrated Project Team Agreement states that:

> The parties and Cluster Partners undertake not to make any claim against each other for any loss or damage whatsoever arising out of or in connection with this Agreement, including claims arising out of prior discussions, claims alleging negligence and claims for injunction relief . . .

Two common alliancing features work particularly well with BIM.

■ In an alliancing project, the parties agree that they will not sue each other, except for willful default. Sharing information cannot lead to liability, and therefore the liability concerns that may impede BIM adoption do not apply in an alliancing project.
■ Because a portion of compensation is tied to a successful outcome, there is an incentive to collaborate. In this context, BIM is an ideal platform for interactively sharing information, ideas and solutions.

A secondary but important theme of alliancing is that people will be motivated to play their part in collaborative management if it is in their commercial and professional interest to do so.

This all leans towards the lean construction approach, which may be distilled into five basic concepts, these being the need to;

■ collaborate
■ increase relatedness between all project participants
■ ensure projects are networks of commitments
■ optimise the project not the constituent pieces of the project
■ tightly couple action with learning.

7.2.1   The BIM model

In addition to all of the above, any BIM contract (whether bipartite or multiparty) will also have to define the status of the BIM model and deal with post-handover matters such as life-cycle management and data capture. A major issue that needs to be addressed is: Does the party managing the model assume additional liabilities and risk?

As noted previously, current practice uses a series of interlocking models to communicate the design and construction intent for a project. In many instances, the complete design is only visualised when imported into a viewing programme. Moreover, most models do not contain all the construction details required for a project. Thus, the contract documents will include some 2D information that is added to the information in the BIM model. In practice, these issues are currently resolved by using a printed submission as the contract document, even if the communication flow has been digital.

However, there are several options currently being followed. The first is that a 'co-contract document' is used between the parties but is not submitted to permitting agencies. In this case, the contracts need to state how inconsistencies will be handled. Another option is to use the BIM model as an 'inferential document'. Under this option, the model provides visualisation of the design intent inferable from the contract documents. Finally, the BIM model can be used as an 'accommodation document' that can be used, but not relied on, by the recipients. This last approach is similar to the CAD transfer liability waivers that designers use when providing CAD documents to contractors. However, limiting reliance in this way undermines the BIM model's utility.

Naturally, the issue of 'ownership of the model' can be worked out through the contract, just as ownership of design documents is now addressed in most standard form documents. However, the issue of ownership becomes much more complex when the final 'model' is actually a gathering of the input of a single model or of many models through the use of software that allows such a roll-up process.

Many parties will have contributed to the 'model' in a fully modelled project, and the issues of design input versus design responsibility will need to be resolved. In addition, the licensing and royalty requirements of potentially 'selfish' members of the building team need to be discouraged in standard form documents. Owners need to be particularly aware of the implications of such issues, and will be expected to play an important role in addressing them.

When a BIM model is used, strict rules must be applied to police the model, so that access rights are reasonably restricted, the ability to change the model is strictly limited to those who are responsible for changes to that portion of the model, outdated versions of the model can be destroyed, and a precise audit trail can be maintained of the various iterations of the model. Anyone who has been involved in a project where one of the participants was working off an obsolete version of the drawings knows that the 2D world has problems of its own. In fact, adherence to 2D drawings as the 'gold standard' of design is built on a faulty premise. The problems long inherent in the use of 2D drawings make clear that regarding them as sacrosanct is a mistake.

Moreover, liability concerns may lead practitioners, and their lawyers, to contractually isolate the BIM model – thus depriving the model of its greatest benefits. Therefore, while the risks presented by BIM may be different in some respects from the traditional route, it does not automatically follow that a contractor's risk should be greater than it is at the present time.

First, contractors should do their best to reach an understanding with all parties about the ability and right to rely on the BIM model. Disclaimers of reliance that some consultants have sought to apply to design documents in electronic format should be discarded.

Second, the same risk allocation principles that apply to traditional 2D design should apply to a BIM model. Even with a BIM approach, the architect/engineer remains responsible under a traditional approach for project design. A contractor's involvement in, and corresponding liability for, design should not extend beyond that typically associated with constructability issues, construction means and methods, and shop drawings. Thus, for example, the fact that the shop drawings are added into a model should not change the risks for the information being added. The crucial question for the contractor is: What are the deliverables and who is responsible for them?

BIM is more than a technology. Although it can be used without collaboration, such use only scratches the surface. Because the model (or models) is a central information resource, it leads naturally to intensive communication and interdependence. BIM models are platforms for collaboration. But collaboration is not a construction industry hallmark. Rather, the industry, its practices and its contract documents assume definite and distinct roles and liabilities. The insurance products used by the construction industry mirror these lines of responsibility and liability. However, collaborative processes, and BIM specifically, foster communication, joint decision-making and interdependence, and these blur the distinctions between parties. Technology and business practices are in collision.

BIM also collides with traditional professional responsibility principles. Although virtually all professional licensing regulations require that designs be prepared by a person 'in responsible charge', much in a collaborative design is not supervised or directed by a single person or entity.

7.3   Integrated project delivery

Collaborative project approaches are beginning to appear in the USA under the general name 'integrated project delivery' (IPD). Unlike alliancing, which has generally been used for major civil and industrial infrastructure, in the USA IPD has been primarily used for complex structures such as hospitals.

IPD is a radical departure from traditional prescriptive and adversarial contract approaches, and offers the potential for increased project value and greater reward for all participants to manage uncertainty and risk, thus eliminating the fear that causes participants to focus on their narrow self-interests. The special-purpose entity contracts with the architect and the contractor to design and construct the project subject to its direction, with incentives and risks based on project performance. This approach is flexible, but complex and, depending on where the project is located, can create corporate governance and licensing issues. The alternative approach is more traditional. The contractor and architect separately contract with the owner under terms that encourage collaboration but still adhere to traditional compensation models, risk allocation and project management roles.

Interlocking risk allocations leave boundaries in place, but lessen their importance. Under this approach, the key participants jointly negotiate specific limitations to their individual liabilities, using releases, indemnifications and limitations of liability. The interlocking risk allocations lessen the liability fears that accompany free flow of information. Interlocking risk allocation has three potential drawbacks.

- there is a risk that the provisions will be inadequately drafted or incomplete
- some jurisdictions have restrictions on liability limitations or indemnification that could undermine this approach
- although the risk allocations lessen disincentives, they do not create any additional incentive to collaborate.

Therefore, to enhance their success, interlocking allocation should be balanced by performance incentives.

7.4   Single-purpose entities

A single-purpose entity is a limited liability enterprise (e.g. corporation, limited liability company, limited liability partnership) created to design, construct and possibly own and operate a facility. Such entities solve the boundary problem by bringing all parties within one boundary.

The key participants sponsor the single-purpose entity and achieve a gain by optimising the entity's success. The single-purpose entity contracts with the owner or employer for the services required to construct the facility, with the specifics of scope, responsibility and liability determined on a project-specific basis. The parties within the boundary must release each other from most potential liabilities, or agree that any 'in boundary' claims will be paid only by project insurance.

Single-purpose entities are common in off-balance-sheet asset financed projects (project finance). Under a classic project finance structure, non-recourse loans are used to design and construct a revenue-generating asset that is owned by the single-purpose entity. The asset, and any guaranteed income streams, secures the loans. As might be expected, there are also numerous variations with limited recourse, limited sponsor guarantees and similar features. However, the fundamental economic principle of the single-purpose entity is that the owner's return is based on creating value in the single-purpose entity.

7.5   Design and build

Design and build contracts (and other similar arrangements) solve the boundary problem between the various design consultants by in effect increasing the boundary's perimeter until it absorbs all the key participants. Thus, information sharing and reliance issues are resolved by joining the provider to the one party. Therefore, in many ways a design and build procurement approach may be seen as being ideal for a BIM approach.

Design and build is a project delivery method that combines two, usually separate, services into a single contract. With design and build procurement, building owners execute a single, fixed-fee contract for both the design services and the construction. For this strategy to be fully effective, the key participants must be identified and included in the design and build team. This is automatically accomplished if the designers are directly employed by the design and build contractor, but it is more challenging if the designers or key systems providers are subcontractors to the design and build contractor. In the latter case, the additional parties can become part of the 'virtual' design and build team if their liability is limited and their compensation, at least in part, is performance based.

With design and build contracts, the design and build company assumes responsibility for the majority of the design work and all construction activities, together with the risks associated with providing these services for a fixed fee. When using design and build delivery, owners usually retain responsibility for financing, operating and maintaining the project.

However, even with design and build procurement the owner must usually complete a certain amount of preliminary engineering and project definition in order to be able to prepare tender documents. A project that is too advanced (e.g. fully designed) may be unattractive, as there will be minimal opportunity for the private sector to apply innovative methods to reduce cost and optimise the schedule. On the other hand, a project that still is at an early stage with unanswered questions regarding scale, alignment and other project features will be difficult to structure on a design and build basis because the potential private sector contractor will be unable to reliably assess schedule and costs. Therefore, even when using the design and build procurement approach there is still the risk of a split of liability with regard to the design functions under a BIM project.

Other variations of the design and build procurement approach are design, build, operate and transfer (DBOT) and design, build, operate and maintain (DBOM).

7.5.1 Design, build, operate and transfer (DBOT)

A DBOT project is typically used to develop a discrete asset rather than a whole network, and is generally entirely new or greenfield in nature (although refurbishment may be involved). In a DBOT project the project company or operator generally obtains its revenues through a fee charged to the utility/government rather than tariffs charged to consumers. A number of projects are called 'concessions', such as toll road projects, which are new build and have a number of similarities to DBOTs. This is a type of arrangement in which the private sector builds an infrastructure project, operates it and eventually transfers ownership of the project to the government. In many instances, the government becomes the firm's only customer, and promises to purchase at least a predetermined amount of the project's output. This ensures that the firm recoups its initial investment within a reasonable time span.

7.5.2 Design, build, operate and maintain (DBOM)

The DBOM approach is an integrated partnership that combines the design and construction responsibilities of design and build procurements with operations and maintenance. These project components are procured from the private sector in a single contract with financing secured by the public sector. With a DBOM contract, a private entity is responsible for design and construction as well as long-term operation and/or maintenance services. The public sector secures the project's financing, and retains the operating revenue risk and any surplus operating revenue. The advantage of the DBOM approach is that it combines responsibility for usually disparate functions – design, construction and maintenance – under a single entity. This allows the private partners to take advantage of a number of efficiencies. The project design can be tailored to the construction equipment and materials that will be used. In addition, the DBOM team is also required to establish a long-term maintenance programme up front, together with estimates of the associated costs. The team's detailed knowledge of the project design and the materials utilised allows it to develop a tailored maintenance plan that anticipates and addresses needs as they occur, thereby reducing the risk that issues will go unnoticed or unattended and deteriorate into much more costly problems.

7.6 How will BIM be used in dispute avoidance/dispute resolution?

As noted above, BIM causes a fundamental change in the way participants in a construction project interact with each other, requiring new and significant levels of collaboration, information sharing and coordination, from the inception of a construction project through to its completion. All participants in the BIM process will be required to sign a common legal document, a BIM protocol that establishes the obligations of the parties, both technical and legal, for the modelling process. The environment created by BIM provides the opportunity for a significant advance in the design of dispute resolution systems for construction projects.

It is generally accepted that a dispute will not exist until a claim is asserted by one party and is rejected or not accepted by the other party. A dispute normally needs the intervention of a neutral body to solve the problem.

Disputes in the construction industry can have a major impact on organisations and projects. The impact of construction disputes can be summarised as follows:

- additional expense in managerial and administration
- possibility of litigation cases
- loss of company reputation
- loss of profitability and perhaps business viability
- time delays and cost overruns
- extended and/or more complex award process
- loss of professional reputation
- reduced respect between parties, deterioration of the relationship and a breakdown in cooperation
- high tender prices
- rework and relocation costs for men, equipment and materials.

Some factors that give rise to dispute in the construction process are:

- failure to plan and schedule adequately
- failure to follow a plan and schedule
- failure to supply adequate manpower
- disagreement over what material is specified
- an inability to agree on what constitutes a variation and the valuation of a variation.

In view of the above, the implementation of BIM presents an opportunity to nip in the bud any disputes before they have the opportunity to develop, and also provides contemporaneous information that should greatly assist in the resolution of a great many other disputes.

For example:

- The ability to view the programme implications of proposed alterations both retrospectively and, perhaps more importantly, as a projection of future actions. If parties have early warning of the potential programme implication of proposed changes, the probability of a dispute arising will be greatly reduced.
- If a dispute regarding programming and/or time matters does arise, it should be easier for a tribunal to resolve such a dispute by referring back to the frozen snapshot views of the programme dimension of the BIM model. While there may still be disagreements about the reason for the delays suffered, at least there should be no dispute about the progress on the project at any particular time. One of the major difficulties in any dispute resolution process is establishing what actually happened and when, and it is plain that the BIM model should greatly assist in this particular issue.
- The required specification for the materials used and to be used will be embedded within the BIM model, and therefore there can be no dispute at a later date about what that specification is. This is in contrast to the present (non-BIM) situation, where there is often a dispute as to whether the material specifications on a drawing, or on a schedule or on a data sheet (where these are different from one another) apply.
- The BIM model contains information, or can link to information, necessary to generate bills of materials, size and area estimates, productivity, materials cost and related estimating information. It avoids the need to process material take-offs manually, and thus reduces error and misunderstanding. Moreover, the linked cost, productivity and resource information evolves in step with the design changes made, and those changes can be recorded as snapshots of the BIM model if required. Therefore, if there is a dispute at a later stage, as BIM technology develops further it is conceivable that the BIM model can be unwound to the point where a problem began to occur, and the source of the problem can then be more easily identified.

The potential of BIM in respect of dispute avoidance and dispute resolution is an almost entirely untapped resource. However, it is clear from the above examples, and in view of the anticipated further developments of the BIM process, that BIM technology may play a major role in both avoiding disputes and resolving disputes.

REFERENCES

BIS (Department of Business, Innovation and Skills) (2011) *BIM. Management for Value, Cost & Carbon Improvement*. Available at: http://www.bimtaskgroup.org/wp-content/uploads/2012/03/BIS-BIM-strategy-Report.pdf (accessed 7 November 2013).

Egan J (1998) *Rethinking Construction*. HMSO, London.

JCT (Joint Contracts Tribunal) (2011) *Fair Payment, Transparency and Building Information Modelling.* Public Sector Supplement. Available at: http://www.jctltd. co.uk/docs/JCT%20Public%20Sector%20Supplement%20Dec2011%281%29.pdf (accessed 7 November 2013).

Latham M (1994) *Constructing the Team.* HMSO, London.

Levene IR (1995) *Efficiency Scrutiny into Construction Procurement by Government.* HMSO, London.

Office of Government Commerce (2007) *Achieving Excellence in Construction.* Available at: http://webarchive.nationalarchives.gov.uk/20110601212617/http://www. ogc.gov.uk/ppm_documents_construction.asp (accessed 7 November 2013).

Procurement/Lean Client Task Group (2012) *Government Construction Strategy.* Final Report to Government. Available at: https://www.gov.uk/government/uploads/ system/uploads/attachment_data/file/61157/Procurement-and-Lean-Client-Group- Final-Report-v2.pdf (accessed 7 November 2013).

Strategic Forum for Construction (2002) *Accelerating Change.* Available at: http:// www.strategicforum.org.uk/pdf/report_sept02.pdf (accessed 7 November 2013).

BIM in Principle and in Practice
ISBN 978-0-7277-5863-7

ICE Publishing: All rights reserved
http://dx.doi.org/10.1680/bimpp.58637.103

Institution of Civil Engineers

publishing

Chapter 8
Insurance, liability and risk

8.1   Insurance

If BIM is used solely to prepare better contract documents, there are few insurance concerns. However, as a collaborative framework, it does create possible issues, and one of the questions often asked is: Do we need new insurance products better tailored to collaborative projects?

Many current professional liability policies for design have exclusions for 'means and methods' and for joint venture liability. The means and methods exclusions are designed to eliminate from design coverage for construction activities. In a collaborative setting, the designers may assist in developing sequences and construction procedures that, at the very least, skirt this exclusion. Sharing risk and reward, a hallmark of integrated project delivery, is also a joint venture characteristic, and may lead insurers to deny or limit liability if joint venture liability is alleged.

Contractors also face insurance issues. Most standard commercial general liability policies exclude professional services and do not cover pure economic losses. As contractors become more deeply embedded in the design process, they must consider whether they should obtain contractor's professional liability coverage. And contractors must also recognise that their standard policies provide little protection from economic claims based on their negligent performance.

Even hosting data can create additional insurance issues. Essentially, data loss more closely relates to valuable paper coverage than traditional construction policies. In addition, if the parties are developing custom software for others' use, there are product risks involved that may not be covered by their usual policies. Software is not perfect, and residual flaws will remain despite strenuous debugging efforts. Luckily, these bugs are most often annoying but not harmful. Sometimes, however, that is not the case. It could be that errors in BIM software cause economic loss to a user but the injured party has no realistic remedy. However, the user's liability to other parties is not similarly limited, and this may cause a liability gap if the errors cause deficiencies in plans or other deliverables.

The insurance industry is aware of these issues and others in respect of the application of BIM, and this may result in an insurance market providing coverage for collaborative projects. However, until that time, participants in the BIM process must work with their insurance brokers to ensure that the tasks they are undertaking on BIM projects are adequately covered by their existing policies.

Few insurance companies currently offer BIM-related products. The rarity of BIM projects to date means that uncertainties remain about its benefits and risks. Consequently, insurers are likely to increasingly offer BIM insurance policies but, until it is clear what risks are involved, premiums are likely to be high. Uncertainty is also likely to impact on the bond market until BIM becomes more commonplace.

Naturally, any convergence of the design and construction processes signals the need for contractors to review not only their overall risk profiles but also their risk-financing programmes. BIM presents many of the same risk management questions contractors already face as they increasingly provide preconstruction services that require them to analyse, price and suggest modifications to the architect's design prior to its completion. Therefore, it is strongly advised that any contractor looking to participate in the BIM process consults with an insurance advisor to examine any potential increase in risk as well as the appropriateness of its current insurance coverage.

Claims arising out of services provided as part of the BIM process may only be passive in nature. That is, they could involve pure economic loss when there is a design error that needs to be corrected. For example, some professional liability insurers may exclude and/ or not specifically address coverage for claims arising from services such as value engineering and constructability review. However, these are services that any contractor may currently be providing without having become involved in the BIM process. Therefore, a thorough review of the contactor's involvement or intended involvement in the design process should be undertaken to assess the potential exposures presented.

Contract documents should be reviewed in conjunction with legal opinion and modified as appropriate, and available insurance protection should be discussed with an insurance advisor prior to entering the realm of potential design risk.

Beyond the design/professional services liability issues that may arise from participation in the BIM process, or that may exist with current project delivery methodologies, the contractor may very well have exposure to other electronic data/technology issues such as project management software, project web hosting/web sharing, transmittal of electronic viruses, or intellectual property ownership/infringement, to name but a few. These potential exposures should also be discussed with legal and insurance advisors to address both the contractual and risk-financing options of these and other technology and/or intellectual property related exposures.

Using BIM substantially alters the relationships between parties, and blends their roles and responsibilities. As we move forward with BIM projects, risks will need to be allocated rationally, based on the benefits a party will be receiving from BIM, the ability of the party to control the risks, and the ability to absorb the risks through insurance or other means.

The risk profile for construction projects and project participants will change with use of a BIM model and a collaborative, integrated approach. Measures will need to be taken to mitigate any increased areas of risk. This may require insurance, indemnities, changes to contracts and/or changes to policies and procedures. A balance will need to be struck between the need for fluid collaboration between parties on the one hand and the need to precisely define responsibility to manage the changed risk profiles on the other.

All in all, the problems of being able to obtain adequate insurance cover for the various participants in the BIM process may be a major factor that impedes the development of BIM.

8.2 Surety bonding industry

The utilisation of BIM technology by the construction industry will be an 'evolving' underwriting process for the surety industry. At present, BIM technology is such a new concept to the surety industry that no clear industry opinion has emerged, and individual sureties are likely still formulating their own positions. This is especially true because little guidance in the form of industry standards exists concerning the proper utilisation and application of BIM technology on construction projects. Some contractors, as early adopters, are utilising BIM technology internally. Such internal use by contractors does not bring any additional burden to the surety, as its use is not made part of the contract documents at present.

However, where contractors utilise BIM technology as part of their contractual performance requirements for a project, certain surety issues may arise, particularly in the absence of a clear contractual delineation between the project parties for design, coordination and communication responsibilities.

Utilisation of BIM technology on projects requires a clear understanding of each party's responsibilities. The parties will need to define and address the data standards and protocols, the preparation of the model and the distribution of information from the model. In this regard, a contractor and its surety which are evaluating BIM should consider how the contractor's organisation can best manage the risks associated with utilising this new project tool. If BIM technology is clearly defined contractually, it should decrease conflicts and make the design and construction process more efficient.

As an example, if a designer approves an electronic file prepared by a detailer, and this file contains a dimensional inaccuracy, the designer must be protected to the same extent that it would have been had the approval document been a printed drawing. Similarly, where a designer provides an analysis model or 3D façade rendering to supplement the design documents, and the designer has stated that the analysis or rendering model shall not be considered to accurately show all aspects of the geometry (presumably the designer has issued and identified a separate document for specification of geometry), the designer shall not be liable if a detailer ignores the warning and bases the geometry on the supplemental model or rendering. The 3D information is valuable and should be shared with the team, but its use and accuracy must be defined carefully.

In summary, it is very early days in attempting to predict what (if any) impact the use of BIM will have on the surety bonding requirements of major construction and civil engineering projects.

BIM in Principle and in Practice
ISBN 978-0-7277-5863-7

ICE Publishing: All rights reserved
http://dx.doi.org/10.1680/bimpp.58637.107

Chapter 9
Example 1 – BIM applied to a large underground railway station

The following case study charts how building information modelling (BIM) may be applied to a large underground railway station and the benefits that may be gained from doing so.

The scope of the works is designed to significantly increase the railway station's capacity, to improve emergency evacuation and to provide step-free access for people with impaired mobility. The upgrade to the railway station will significantly increase the size of the existing ticket hall, and a new entrance and a further ticket hall will be created. This new ticket hall will be linked to the platform tunnels, to the existing ticket hall and to the platforms by means of new passenger tunnels. New lifts will provide step-free access from street to platform levels. The new tunnels will be approximately 6 m in diameter; ticket hall boxes will be up to 10 m deep. Both the existing and the planned new ticket hall will be under busy public roadways, and because of this, and to minimise traffic disruption, the works will principally be carried out using a top-down construction. The new ticket halls and tunnels will be built adjacent to existing infrastructure and, being in an inner city location, the site is criss-crossed by existing pipelines, sewers and power and communications cables.

In addition, construction will take place adjacent to or beneath the foundations of adjacent existing buildings. Piles will be used to form the walls of the new ticket hall, and those piles will be within 5 m of an underground culvert carrying a watercourse. There is also limited clearance between the passenger tunnels linking the proposed new ticket hall with the existing platform tunnel. Space is restricted, and it is imperative that construction activities do not cause undue disruption.

All in all, the logistics of carrying out the works is extremely challenging. To manage this somewhat daunting task and the risks associated with the proposed underground railway station upgrade, the project team proposes to use BIM. BIM is a process that incorporates 3D design, simulation and analysis, quantity surveying and a host of other tools, and

107

provides a base or platform for collaboration. It has been decided to use the BIM approach because, without a spatially accurate, fully coordinated 3D model it would be virtually impossible to visualise and coordinate the complexities of the project and the surrounding infrastructure, services and other 'obstructions' and limiting restrictions.

To make most use of BIM, it has been decided that 15 discrete design disciplines will be concentrated on as individual BIM models, showing how the entire project fits together as one composite BIM model.

The BIM working process is built around the concepts of:

- collaboration between the client and a federated project supply chain
- a single, unified system for data creation, management and sharing
- a coordinated information model

To maximise interoperability between disciplines, one standard software package has been chosen, and this will be incorporated in the BIM protocol as the standard software to be used by all BIM participants.

To coordinate the new planned infrastructure with the existing infrastructure, and to minimise the risk of clashes, it is planned to create a 3D record of the existing infrastructure and the other obstructions and services. This 3D record will be based on 'as-built' records, where available, supplemented by data generated by millimetre-accurate laser surveys. When work commences, this existing as-built data will be supplemented by logging onto the BIM model the exact location of cables, pipelines and other structures encountered.

The BIM model will also be used to ensure that the proposed piled walls are raked in order to avoid the services and other obstructions that may be encountered, and that when constructed the piles leave no voids behind when excavation work commences. This information will also be used to ensure that the piling rigs are positioned in the correct position to suit the raking required for the piles forming the walls. Through taking this approach, there will be an improvement in accuracy and efficiency, and the risk of clashes etc. occurring will be mitigated.

The BIM model(s) will be used to check for structural, architectural and building services clashes within the railway station. If any clashes are encountered and redesigns of the structure are necessary to avoid the obstructions, the BIM model(s) will update automatically to provide the required structural calculations and the revised size of structural columns and beams etc. The same approach will also be used where, because of architectural changes for example, the structure needs to be altered to suit. In such a case, the

Example 1 – BIM applied to a large underground railway station

BIM model(s) will automatically make the structural calculations allowing for the revised load paths, and will regenerate the revised reinforcement design with minimal rework.

As quantities of materials will be automatically calculated by the model, this will provide immediate details of the effect on cost of such a change and the time implications on the programme for the works.

3D printing technology will be used to generate physical scale models of the virtual BIM model. In combination with viewing tools such as interactive PDFs, this will significantly improve the project team's ability to communicate their design and construction intentions between themselves and to stakeholders and to other interested parties. This will contribute to the timely feedback and modifications required for any other reason, which will avoid the potential cost and delay associated with later stage changes.

As the project moves into the build phase, the richness of information contained within the 3D BIM model(s) will improve the contractor's ability to understand and resolve details. It will also enable the project team to check work against specifications and validate any anomalies.

The integrated single BIM model will enable continuous review and permit testing of the design for constructability and for the maintenance and operation of the railway station.

Post-completion, the detailed model will provide the building owner with an accurate record of assets – those belonging to third parties as well as its own assets. Containing spatial and technical information, the integrated BIM model has the potential to assist the building owner in managing the railway station throughout its life cycle.

BIM in Principle and in Practice
ISBN 978-0-7277-5863-7

ICE Publishing: All rights reserved
http://dx.doi.org/10.1680/bimpp.58637.111

Institution of Civil Engineers

publishing

Chapter 10
Example 2 – BIM applied to precast concrete fabricators

Sophisticated 3D parametric modelling software for precast concrete structures has been commercially available since 2005. This example demonstrates how participants in the precast concrete field would likely have gained sufficient experience with building information modelling (BIM), as applied to precast concrete buildings, to allow analysis and comparison of that approach with their normal pre-BIM approach.

The example sheds light on the obstacles that would have needed to be overcome, the achievements and the disappointments in respect of the application of BIM, and the changes in workflow and personnel that would have been experienced. The example may help other potential participants in BIM to avoid some of the pitfalls of replacing 2D computer aided design (CAD) practices with BIM.

By way of background, extensive research and development in academia and industry led to the emergence of powerful and practical BIM tools for structural analysis, design and detailing. These tools were gradually being adopted by specialist precast concrete companies. However, software tools alone were insufficient for successful BIM adoption. Deep changes in terms of work practices, human resources, skills, relationships with clients and contractual arrangements were required for success. Indeed, it was accepted that the best results could only be achieved when change extended beyond the borders of any individual organisation adopting the new BIM technology.

The early adopters of BIM tools in precast concrete production were among the first pioneers in this area, and a consideration of the how they would have approached things provides the opportunity to research what practices work, how adoption of BIM can be pursued and what the impacts of BIM are.

Precast concrete is a construction method in which concrete is cast in reusable moulds and cured in a controlled environment, then transported to the construction site and lifted and fixed in the structure. The two main types of use are for structural elements

111

(such as beams, columns and slabs, which may or may not be prestressed) and for architectural façades.

There are five main participants in the precast construction process: the building owner, the architect, the structural engineer, the precast concrete designer/fabricator and the precast concrete erection crews. Some precast concrete designer/fabricators maintain engineering staff in house, while others procure the service from independent engineering design firms. In this example, the precast concrete designer/fabricators maintained engineering staff in house.

BIM research and development for the architecture, engineering and construction industry in general focuses on provision of parametric 3D modelling software and on achieving interoperability between various applications.

The directly measurable benefits of BIM for the precast concrete industry were expected to be significantly reduced engineering costs and significantly reduced costs of rework due to errors. In addition, and potentially more significant, benefits were expected to arise from enhanced cost estimating accuracy, a great reduction in engineering lead time, an improved customer service and a boost for automation in production.

Naturally, as BIM represents a paradigm shift from the use of 2D computer-aided drafting, the company concerned in this example anticipated that the transition was likely to involve personnel issues. It also considered that the shift was likely to present the opportunity for rethinking and possibly re-engineering existing workflows and information flows in both engineering and in production. Because of this, before using the BIM technology, the company concerned carefully prepared strategies and working plans for the adoption phase of BIM, and implemented monitoring procedures to enable a benchmarking process.

Being a typical precast concrete designer/fabricator, the company realised that there would be benefits in preparing precast concrete shop drawings using the BIM concept. The idea of 'pre-building' a precast project in the virtual world of BIM software, and ensuring all geometry, details and connections within the model were correctly placed and coordinated was seen as being extremely useful to reduce the likelihood of errors.

During fabrication or erection of precast elements, geometry or connection errors on drawings can be costly to repair, so traditionally much time is spent on checking and vetting drawings. Technologists preparing traditional CAD drawings of precast elements must use the 2D drawings and visualise the structure in 3D, and coordinate between numerous drawings to ensure that nothing is overlooked. The company in this example realised that, as a first step, BIM could be used to review potential conflicts or project

Example 2 – BIM applied to precast concrete fabricators

complications within the model, so they could be easily discovered and resolved prior to issuing drawings for construction. The use of the BIM software would also inherently reduce the possibility of misaligned connections, incorrect architectural features and geometry conflicts, so that shop drawings could be created without the need for detailed checking or cross coordination between drawings.

As a specialist designer/contractor, the long-term goal of the company in question was to use the BIM technology to enhance productivity and quality in producing designs and drawings. This was important because in the design element of the works labour is the major resource, with drawing and checking consuming in excess of 80% of the labour input in typical projects. While, reducing the likelihood and frequency of drawing errors was a short-term goal, the company's management realised that, even if there was no immediate reduction in man-hours by moving from CAD drafting to BIM, the benefits of reducing design and drawing errors would be a key advantage.

In addition, the company realised the benefits that BIM would bring in providing a database of information that would be useful for building owners and other interested parties. As the company already used computer design and analysis software, it realised the future benefit of having BIM software that could harmonise with analysis software and streamline the engineering process.

The other major benefit that was foreseen was the ability to absorb more easily design changes initiated by building owners or main contractors.

During the several years preceding the adoption of BIM, the company had provided precast concrete design and units for several major projects. Difficulties related to design and drafting errors that led to problems of mismatched pieces and connections in the field, low productivity in preparing shop drawings (especially where design changes were frequent), and long cycle times for design reviews led the company's principal engineers to consider BIM as a means of improving their precast concrete design and supply service.

The company established four main objectives:

- to increase its capability to *absorb design changes* with a minimum of rework in preparing and reconciling different drawings
- to harness the capabilities of 3D visualisation of the project in order to check for and avoid design errors (this was particularly necessary for viewing complex and congested arrangements of embeds, reinforcing and prestressed strands)
- to improve productivity, by producing schedules and shop drawings for precast structures in as automated a fashion as could be achieved

■ to visualise the structure, specifically to show the owner the spatial precast elements in 3D.

To consider and monitor the progress achieved along the learning curve of BIM adoption, the company first carried out a case study project. The project entailed the design of precast façade panels with complex piece geometries.

On the chosen project, the geometry was complex. The building's walls were stepped and curved both horizontally and vertically. The new architectural façade was to be cast as numerous flat panels, but the pieces had to follow the existing curved structure. Precast pieces included wall panels notched around existing windows, spandrel panels, column covers and base cladding panels. It was critical that the panels' geometry should be aligned with the openings, doors and windows in the existing cast-in-place concrete and steel walls. All the pieces had diagonal intersecting reveal patterns. As a result, the company elected to make its first exclusive use of BIM the architectural façades on this project, primarily to ensure proper and accurate geometry.

First, a digital survey was undertaken on site, and the existing building geometry and all openings were provided in a 3D CAD file. The CAD file and the architectural CAD floor plans were imported into the BIM software, and the precast panel geometry was developed to match the 3D survey.

During development of the precast model, it was discovered that additional as-built survey information was required to provide geometry in complex areas, so additional digital survey information was requested. This was provided by the main contractor and imported into the model to complete the modelling. All the erection layout drawings and the individual precast panel drawings were created using the BIM software.

As this was the company's first use of the BIM software, and as the work related to architectural panels, the drawings were exported to CAD for final touch-up and to add the manufacturer's standard lift hook details. Steel connection hardware drawings were issued as CAD drawings using the precast manufacturer's standard hardware drawings.

The entire project was modelled and designed by a single engineer, who had training and was familiar with the software features and capabilities, but who had not used it on a real project. The total labour hours recorded was about the same as the estimate of the input that would have been required to complete the project using traditional CAD procedures. However, there were no drawing errors that led to construction problems on site, and all pieces built in accordance with the shop drawings fit the complex curved geometry of the structure. All of the reveals and architectural features lined up between adjacent pieces.

Example 2 – BIM applied to precast concrete fabricators

As a result of the issues discovered during modelling, the architectural drawings were updated and revised to suit decisions made from viewing the BIM model.

The engineering calculations were undertaken by an intermediate level engineer, with some additional engineering provided by a senior engineer. Design software was used for much of the analysis and design.

On a typical CAD precast project of this magnitude, a team of at least three drafting technologists would have been used to create and check all the drawings. The project was modelled and issued on schedule with just two modelling personnel: an engineer who was familiar with BIM and a junior technologist who was new to the software. This was the firm's first experience using BIM for a complete precast structure.

All erection layout drawings and all piece cast unit drawings were created and edited within the BIM software, and issued as PDF files to the precaster.

Once the model was complete, the shop drawings were created automatically from the software. The time needed to edit and complete each drawing ranged from 5 minutes to 2 hours (depending on the complexity of the piece, which ranged from simple hollow-core panels to complex beams or wall panels). Some texts needed adjusting to avoid overlap, and standard CAD lift hook or other manufacturer standard details were imported into the BIM drawings. For comparison, typical durations for CAD drafting of typical wall panels, columns or beams would normally have been 4–8 hours per drawing.

Temporary features required for the fabrication, transport and erection of the precast pieces, which were not part of the final building, were also modelled and shown on the relevant drawings. These included lifting steel 'strongbacks' bolted to the back of some panels with large openings (required to strengthen the panels as they are lifted out of the forms, and to secure and stabilise those particular panels during transport to site) and temporary bracing (needed for stability during erection).

After the first structure was erected, changes were requested by the architect and the building owner, including revising wall-panel reinforcement from rebar to mesh, moving and adjusting the sizes of mechanical blockouts, revising some lift hooks, and minor changes to some connections to make erection simpler for site personnel. All the changes were made in the model and drawings were updated automatically.

Erection layouts were simplified, as only the information required for the erection personnel was needed on the drawings. This included geometric layout and connection details. Overall, the amount of dimensioning and information presented on the erection layouts was substantially reduced. Previously, using CAD, a host of details had to be provided on the layout drawings in order to provide enough information for drawing the individual precast pieces, but this information was not relevant to the erectors and was not required using the BIM procedure.

An important result of this was that, throughout the erection process, no repairs were required due to errors relating to the shop drawings. This was considered a major achievement by the company carrying out the case study.

The company's adoption strategy was predicated on its assumption that the long-term commercial benefits of using BIM were clear, and therefore that being an early adopter of BIM would give the firm a competitive advantage. Because of this, there was strong leadership and a commitment to overcome any obstacles in the use of BIM through persistence. This attitude was reflected in the company's approach to training and the willingness to work around the problems presented by early bugs in the BIM software.

The company decided to develop its adoption, in terms of gaining skills, in four main stages:

1 basic 3D modelling
2 automation of drawing production
3 preparation and use of sophisticated parametric components
4 use of integrated structural analysis functions.

From the start, junior drafting technicians were selected for training rather than experienced drafters or engineers, on the assumption that they would have less difficulty in 'un-learning' work patterns suited to 2D CAD. This was based on the understanding that the workflows best suited to BIM would be different from those that had evolved for CAD. The training was aligned with this overall strategy.

A careful analysis was made of what functionality was essential for basic modelling and drawing production, and only those aspects were taught. Training for, preparation and use of sophisticated parametric components was postponed until after the first case study project had been completed. All the training was undertaken in-house by the principal engineer, who had the benefit of extensive prior experience with the software.

The conviction of the company's leadership in the viability of BIM led it to apply the process to actual projects from the start, without ghosting projects that were already underway using CAD. This 'sink or swim' approach forced the company to rapidly develop skills and working methods. The company's decision to proceed in this way was bolstered by the support it received from a major client (a large contractor) that shared their vision and understood the benefits that the adoption of BIM would bring to its own

Example 2 – BIM applied to precast concrete fabricators

business (the first benefit it wanted to achieve was short lead times for the provision of shop drawings, which could not be achieved with CAD). The client's management was willing to accept that there would be teething problems and was prepared to guarantee a flow of work while those teething problems were ironed out. Indeed, a strongly symbiotic relationship arose between the company and the client because the client was keen to prevent the company offering its new expertise to the its competitors.

The main problems encountered by the company during the BIM adoption process were software bugs and the use of incorrect modelling procedures. Both reduced productivity and resulted in much rework.

The lack of formal training and effective online support meant that the firm honed its modelling methods through trial and error before an optimal approach was formed. Difficulties with logical numbering of precast pieces for production and producing bills of material were common because the software's ability to identify like pieces for production is sensitive to the way in which they are modelled. In retrospect, the company considered that focused advanced formal training may have alleviated some of those problems.

The shortage of skilled operators was identified as a threat, and so trainees were initially asked to commit in writing to remain with the company for a fixed period of time before being trained in BIM.

During the case study project, during which emphasis was placed on modelling alone, custom components were only created on an ad hoc basis, as they were needed for that particular project, and it was not considered to be worthwhile building an extensive library of components.

Three significant aspects in the application of BIM were noted.

- The sequence of preparing precast drawings in 2D CAD involves paying attention sequentially to general arrangement plans, then cross-sections, then piece views and finally rebar bending schedules. Much effort is required to coordinate between the different pieces. In BIM, the approach is holistic, with attention paid to a single model, in 3D.
- The BIM system was able to produce a detailed bending schedule automatically, which meant that the rebar callout tag on the section could contain just the rebar mark number. It was therefore no longer necessary to provide all the information about each rebar within the tag on the particular component drawing. Nevertheless, the engineer concerned had invested significant time and effort 'forcing' the system to generate the detailed rebar tag; and it was only much later,

after discussion with site engineers who were more familiar with the BIM technology, that this was found to be unnecessary.

- The final part of the adoption process was to compile templates for 2D drawings and reports according to the company's needs. Here too, much time was invested in attempting to achieve templates that were as close as possible in style and content to those traditionally used by the company. However, in the final analysis, it was generally agreed that this was unnecessary, as many of the standard templates provided with the BIM software made better use of the 3D modelling paradigm.

In many ways, this firm gained experience in BIM the hard way, which is inevitable to a degree for any pioneer. Notwithstanding this, the company cited engineering productivity and improved quality of design and documentation as primary motivating factors behind its move to BIM. The erection crews reported zero or negligible erection delays and waste resulting from design errors. However, the goal of enhanced productivity takes more time to achieve, and the learning curve for the company was steep.

It became evident to the company that the levels of productivity reached and the pace of productivity gain were highly dependent on the degree of formal training provided. Formal training and less formal ghosting and acclimatisation periods appear to be essential.

The company reported that its BIM operators needed to undergo a significant change in thinking in moving from the CAD approach to precast concrete engineering design to a BIM approach.

Another point that the company picked up was that the productive use of BIM requires careful planning of precisely how a building is to be modelled, which is a level of sophistication unnecessary for CAD operation. In addition, the choice of objects used, the way in which parameters are applied to custom components, the way in which elements and rebars are aggregated within details and applied to precast pieces, and the way in which parametric connections are modelled and applied between pieces all have strong impact on the ways in which drawings appear, the level of automation that can be achieved and the types of material reports that can be obtained. A corollary is that professional support is of cardinal importance during and immediately after the adoption phase.

A main goal of the company was the to shorten the time required for preparation of precast engineering documentation, and this was seen as being a key criterion for winning future projects. The company saw that the adoption of BIM gave it a clear commercial advantage over its competitors, and it also gave it the ability to absorb changes late in the process and produce accurate shop drawings rapidly with minimal rework.

Example 2 – BIM applied to precast concrete fabricators

However, exploitation of this advantage was, in some cases, limited to the degree to which the drawings could be produced automatically, and the company encountered limitations in this regard with the versions of software used in the case study, which resulted in the need to manipulate drawings 'manually'.

The company had focused its initial use of BIM on a single engineer operator, and this rapidly became a limiting factor. Based on that experience, it is clear that dependence on a single individual is to be avoided, not only due to the risk it entails, but because all but the smallest projects require more than one operator to achieve the goal of shortening the project duration.

The case study supports the hypothesis that the workflow on a precast project designed using BIM is different to that on one designed using CAD. The main difference is in the change in focus of the modeller/drafter. For much of the BIM workflow, the focus is placed on the building as a whole, with all the work performed on the model. Drawings are secondary. This is in contrast to CAD, where all the work must be performed on the drawings, and the whole building is only modelled in the designers' minds. As such, design using BIM becomes largely top-down, as opposed to a hybrid, iterative approach when using CAD.

A major part of the work in BIM is to create parametric libraries of details, connections and objects so that modelling can be made efficient and to ensure geometric compatibility between adjacent pieces. The modeller concentrates more on the design and engineering aspects of the project and less on document production.

The company in this example had no request from the building owner for BIM data for facility management purposes, and thus drew the conclusion that the facility management industry is not yet sufficiently aware of what BIM can offer.

The company undertook a strengths–weaknesses–opportunities–threats (SWOT) analysis when considering adopting BIM, and this may be a useful guide for other potential participants in BIM.

Strengths	Weaknesses
Skilled engineering staff experienced in CAD and other software	Skilled operators are in short supply and are costly to train
Appropriate IT infrastructure, access to advanced software	Adoption requires capital investment
Leadership with vision	

Opportunities	Threats
Increased engineering productivity	Varying workloads
Enhanced competitiveness of engineering services through reduced design lead times and virtual elimination of geometry and design consistency errors	Dependence on a small number of engineers skilled in BIM
	Staff who are unable or unwilling to adapt may feel threatened
Provision of new services for owners and contractors (e.g. visualisation for conceptual design, rapid and accurate quantity take-off and estimating, data for monitoring and managing production and erection)	Drawings cannot be produced fully automatically: 'manual' editing is still needed
	Inability to remain profitable without BIM if competitors adopt

In conclusion, it is clear from the case study that the adoption of BIM is challenging and not all of the hoped for goals may be achieved in its first use. BIM is likely to become more effective the more often that it is used.

However, it does appear that an improved quality of engineering information can be achieved, and BIM unequivocally improves the quality of engineering design in terms of accuracy and reliability of the documents. Fabrication and erection are essentially error free, and the time required to check drawings is reduced drastically.

It is clear that the adoption of BIM needs to be carried out in carefully measured stages. It is sensible to gain basic modelling and drawing production skills before progressing to productivity-enhancing functions such as the use of parametric custom components. Structural analysis using BIM data needs to be considered to be a medium- to long-term goal.

It is also clear that BIM is a powerful but complex technology. To make progress in adoption, a company needs to establish and maintain its organisational knowledge. While formal training by, or in consultation with, an expert experienced in precast concrete can avoid incurring the costs of low productivity and rework as this knowledge is developed, the procedures really need to be developed by internal staff.

Leadership by management is critical in the early stages, where human resource issues arise and frustration may be felt. For example, BIM tools can be exploited best in work-flows that are different from those that have been evolved for work with paper drawings and CAD tools. With BIM, documentation is produced rapidly and cheaply, so multiple

Example 2 – BIM applied to precast concrete fabricators

reports and drawings can be prepared that provide the exact information needs of the various end users.

Therefore, in summary, **BIM** can be of great benefit to a company, but there are significant hurdles that need to be overcome along the way.

BIM in Principle and in Practice
ISBN 978-0-7277-5863-7

ICE Publishing: All rights reserved
http://dx.doi.org/10.1680/bimpp.58637.123

Institution of Civil Engineers

publishing

Chapter 11
Example 3 – A sample COBie spreadsheet

Today, most contracts require the handover of paper documents containing equipment lists, product data sheets, warranties, spare-part lists, preventive maintenance schedules and other information. This information is essential to support the operations, maintenance and management of the facility's assets by the building owner and/or property manager.

Gathering this information at the end of the job, today's standard practice, is expensive, because most of the information has to be recreated from information created earlier. COBie simplifies the work required to capture and record project handover data. The COBie approach is to enter the data as they is created during design, construction and commissioning.

In a typical construction project the information about the building is contained in drawings, bills of quantities and specifications. A number of construction professionals normally collaborate to put this documentation together. The documentation should then be updated throughout the construction phase, and handed to the client on completion of the project. In reality, this does not always happen, or when it does the documentation is supplied in a format that may be difficult for the client or the building owner to use.

The idea behind COBie is that the key information is all pulled into one format and shared between the construction team at defined stages in a project.

A COBie spreadsheet is by no means a full BIM, but it does contain structured content from all members of the construction team and from many information models. The tables below show one of the worksheets that defines the floors, spaces and zones which make up the building. Spaces can exist on floors. Groups of spaces can also be categorised in zones. This information would typically be extracted from the CAD model. Table 11.1 is an example of a 'floor' worksheet.

Table 11.1 The 'floor' worksheet within a typical COBie spreadsheet

Name	Created by	Created on	Category	External system	External identifier	Description
Floor 1	ABC Co.	10-04-2013	Floor	Auto Desk	C/001/F1	Floor 1
Floor 2	ABC Co.	18-04-2013	Floor	Auto Desk	C/001/F2	Floor 2
Floor 3	ABC Co.	23-04-2013	Floor	Auto Desk	C/001/F3	Floor 3
Roof	ABC Co.	30-04-2013	Roof	Auto Desk	C/001/F4	Roof

Table 11.2 The 'type' worksheet within a typical COBie spreadsheet.

Name	Asset type	Manufacturer	Model number	Warranty/ guarantee	Warranty duration	External identifier
Bath Type A	Fixed	BCD Co.	BH/001/A	W	12 months	BH/001/A
Bath Type B	Fixed	BCD Co.	BH/001/B	W	12 months	BH/001/B
Basin Type A	Fixed	BCD Co.	BN/001/A	W	12 months	BN/001/A
Basin Type B	Fixed	BCD Co.	BN/001/B	W	12 months	BN/001/B

Table 11.2 is an example of a 'type' worksheet. This information would usually be extracted from the specification, and the worksheet lists the types of products in the building, their reference and contact information for the manufacturer. Not shown is the 'attribute' worksheet, which details the properties for each of the type objects. An example of this would be the colour and the material of the bath item and any standards it conforms to.

Each instance of these type objects are then detailed in the 'component' worksheet (Table 11.3). In the table the Bath instance BH/001/A is in space Floor 1. This information would come initially from the CAD model and would then be linked to the type information from the specification model.

The 'contact' worksheet (Table 11.4) lists contact details, company information and company roles. This worksheet list a wide range of contacts, including the design

Table 11.3 The 'component' worksheet within a typical COBie spreadsheet

Name	Created by	Created on	Type name	Space	Notes
BH/001/A	ABC Co.	10-04-2013	Bath Type A	Floor 1	Description to add
BH/001/B	ABC Co.	10-04-2013	Bath Type B	Floor 1	Description to add
BN/001/A	ABC Co.	10-04-2013	Basin Type A	Floor 1	Description to add
BN/001/B	ABC Co.	10-04-2013	Basin Type B	Floor 1	Description to add

Example 3 – A sample COBie spreadsheet

Table 11.4 The 'contact' worksheet within a COBie spreadsheet

Email	Company	Phone	Department	Function/components
ed@abc.com	ABC Co.	0208 111 222	Design	Design
jo@abc.com	ABC Co.	0208 111 222	Drafting	Design
bob@bcd.co.uk	BCD Co.	0207 222 333	Sales	Sanitaryware
tom@bcd.co.uk	BCD Co.	0207 222 333	Supplies	Sanitaryware

Table 11.5. The 'job' worksheet within a typical COBie spreadsheet

Name	Created by	Created on	Category	Status	Type name
Annual inspection of doors	CDE Co.	10-06-2013	FM	Not yet started	Doorsets
Monthly fire alarm test	CDE Co.	10-06-2013	FM	Not yet started	Fire alarm
Quarterly intruder alarm test	CDE Co.	10-06-2013	FM	Not yet started	Intruder alarm
Annual fire alarm inspection	CDE Co.	10–06-2013	FM	Not yet started	Fire alarm

consultants, the construction team, and the manufacturers and suppliers of the systems and products. This information would typically come from the specification and preliminary documents.

The spreadsheet also contains worksheets such as a 'job' worksheet (Table 11.5), which allow specific facility management tasks to be assigned to the various objects in the building.

Within the spreadsheet it is possible to link the data. So, for example, the Bath instances can be linked to the specification of the bath type through a simple drop-down box. It is through this interlinking (in line with the example tables shown) that a unified set of data about a partial BIM model can be fully utilised.

BIM in Principle and in Practice
ISBN 978-0-7277-5863-7

ICE Publishing: All rights reserved
http://dx.doi.org/10.1680/bimpp.58637.127

Chapter 12
Conclusion

Currently in the UK, there is a lot of discussion about building information modelling (BIM). BIM is a way of approaching the design and documentation of a project utilising 3D computer technology that is shared between the design and construction teams, incorporating cost, programme, design, physical performance and other information regarding the entire life cycle of the building in the construction information/building model.

In the UK this discussion has largely been generated by the publication of the UK Government's construction strategy (BIS, 2011), which requires that all government projects utilise BIM in the form of a fully collaborative 3D computer model (Level 2) by 2016, with all project and asset information, documentation and data being electronic.

Although knowledge of BIM is growing rapidly in the UK, its use presently lags far behind that in other countries. Internationally, the use of BIM has already been used on projects worldwide. For example, BIM is used widely in the USA, and in Norway the Statsbygg (the Norwegian Government's key department involved with construction and property affairs) already uses BIM on all public projects.

As the use of BIM accelerates within the design and construction industry in the UK, it will inevitably lead to a revolution in project delivery. Many are already referring to this revolution as leading to the fully collaborative project team. The theory of full collaboration generally envisages the entire project team – owner, architect, engineers, consultants, contractor and specialty contractors – being involved from the inception of a project, by 'sitting together at one table' in developing the project design. Essentially, the team constructs the project electronically in 3D with the use of BIM. This full collaboration allows for increased speed of project delivery, enhanced economics for the project and true lean construction all at levels – something that has never yet been experienced in the UK construction and civil engineering industries.

Building owners are under increasing pressure to deliver projects more efficiently and expeditiously due to more intense worldwide competition within their own industries.

Increased or full collaboration fulfils the building owner's new and future objectives in delivering projects.

We must face the fact that the UK design and construction industry has historically reacted rather slowly to change, especially revolutionary change. However, the use of BIM, for those who react quickly, has the potential to propel them beyond their competition in great strides.

For contractors, any use of BIM will improve the way in which they conduct their business, and they do not need to wait until other specific areas of the construction and civil engineering industries adapt to the use of BIM. Many specialist contractors and subcontractors are already well versed in 3D coordination, and with the recent developments in BIM the software tools are now available for contractors to coordinate the overall project by using 3D models, thus allowing incorporation of the related efforts of 3D design by specialist subcontractors. Simply put, contractors can move further ahead simply by coordinating a project in 3D as opposed to utilising existing 2D methods. The reduction in field corrections and re-fabrication alone provides tremendous increases in labour production and resultant cost savings.

As stated in the introduction, BIM is a tool. Recent advances in computer hardware and software have made BIM technology available and relevant to the work of all members of a project team. The use of BIM may well change the ways in which projects are conceived, designed, communicated and defined, but this tool will not change the core responsibilities of the members of the project team.

In a fully integrated 3D virtual construction environment contractors and construction managers will still need to organise and lead the onsite construction effort. No amount of technology will replace the need for a well thought out approach to construction that will allow each contractor and specialist subcontractor to apply its skills in a safe environment.

Similarly, BIM will not replace the need for designers to convey their design intent, nor will it replace the dialogue of the submittal process through which contractors and subcontractors demonstrate their interpretation and understanding of the design intent.

It is clear that the digital revolution around us is influencing our everyday lives, through the way we access movies, books and music, and even in the way we search for a new house or car. This technology and process change is now also revolutionising the construction and civil engineering process.

When considering BIM, there are potentially major advantages to be gained – generating drawings from a single model, clash detection, creation of visualisations, to name just a few. However, to get to that next level a great deal of effort is needed to bring BIM close to the top of the agenda, with a focus on the well-structured information that can be fed into the models. Information is at the heart of BIM, and standardised, well-structured information will enable enormous efficiencies in the construction and civil engineering industries.

BIM promises exponential improvements in construction quality and efficiency. However, current business and contract models do not encourage, but rather tend to inhibit, the application of collaboration. Therefore, to bring BIM into the mainstream, business models and contract relationships need to be re-crafted to reward 'best for project' decision-making rather than 'best for individual participant' decision-making, and to allocate responsibility equitably between all construction participants.

BIM is the future, and because of that, while it needs to be approached cautiously and wisely, it needs to be embraced with hope and optimism as a further major technological step forward for the construction and civil engineering industries.

REFERENCE

BIS (2011) https://www.gov.uk/government/publications/government-construction-strategy (accessed 6 December 2013).

BIM in Principle and in Practice
ISBN 978-0-7277-5863-7

ICE Publishing: All rights reserved
http://dx.doi.org/10.1680/bimpp.58637.131

Index